CAMBRIDGE INTERNATIONAL A/AS-LEVEL

BIOLOG

C000062932

REVISION GUIDE

Mary Jones

HODDER
EDUCATION

Hachette UK's policy is to use papers that are natural, renewable and recyclable products and made from wood grown in sustainable forests. The logging and manufacturing processes are expected to conform to the environmental regulations of the country of origin.

Orders: please contact Bookpoint Ltd, 130 Milton Park, Abingdon, Oxon OX14 4SB. tel: (44) 01235 827827; fax: (44) 01235 400401; email: education@bookpoint.co.uk. Lines are open 9.00–5.00, Monday to Saturday, with a 24-hour message answering service. Visit our website at www.hoddereducation.co.uk

Impression number   7
Year   2014

Illustrations by Greenhill Wood Studios
Typeset in ITC Leawood 8.25 pt by Greenhill Wood Studios
Printed by CPI Group(UK) Ltd, Croydon

A catalogue record for this title is available from the British Library

ISBN 978 1 4441 1267 2

Contents

Introduction

■ ■ ■

AS Content Guidance

■ ■ ■

AS Experimental Skills & Investigations

■ ■ ■

AS Questions & Answers

■ ■ ■

A2 Content Guidance

■ ■ ■

A2 Experimental Skills & Investigations

■ ■ ■

A2 Questions & Answers

introduction

Introduction

About this guide

This book is intended to help you prepare for your Cambridge International A & AS Level Biology examination. It is a revision guide, which you can use alongside your usual textbook as you work through your course, and towards the end when you are revising for your examination.

The guide is divided into two main parts. Pages 13 to 136 cover the AS examination. Pages 137 to 256 cover the A2 examination.

- This **Introduction** contains an overview of the AS and A2 Biology courses and how they are assessed, some advice on revision and advice on taking the examination papers.
- The **Content Guidance** sections provide a summary of the facts and concepts that you need to know for the AS or A2 Biology examination.
- The **Experimental skills** sections explain the practical skills that you will need in order to do well in the AS Advanced Practical Skills examination, and in the A2 Planning, Analysis and Evaluation examination.
- The **Questions and Answers** sections contain a specimen examination paper for you to try. There are also two sets of students' answers for each question, with comments from an examiner.

It is entirely up to you how you use this book. We suggest you start by reading through this Introduction, which will give you some suggestions about how you can improve your knowledge and skills in biology and about some good ways of revising. It also gives you pointers about how to do well in the examination. The Content Guidance will be especially useful when you are revising, as will the Questions and Answers.

The syllabus

It is a good idea to have your own copy of the Cambridge International A & AS Level Biology Syllabus. You can download it from:

http://www.cie.org.uk/qualifications/academic/uppersec/alevel/

The **Syllabus Content** provides details of the biological facts and concepts that you need to know, so it is worth keeping a check on this as you work through your course. The AS Syllabus Content is divided into eleven sections, A to K. The A2 Syllabus Content is divided into ten sections, L to U. Each section contains many learning outcomes. If you feel that you have not covered a particular learning outcome, or if you feel that you do not understand something, it is a good idea to work to correct this at an early stage. Don't wait until revision time!

Do look through all the other sections of the syllabus as well. There is a useful section on **Definitions**. As you work through your course, you could use a highlighter to mark each of these definitions that are relevant for the topics you have covered. It is worth learning each one thoroughly.

Syllabus content

The content of the AS syllabus is divided into 11 sections:

A Cell Structure
B Biological Molecules
C Enzymes
D Cell Membranes and Transport
E Cell and Nuclear Division
F Genetic Control
G Transport
H Gas Exchange
I Infectious Disease
J Immunity
K Ecology

The content of the A2 syllabus is divided into 10 sections:

L Energy and Respiration
M Photosynthesis
N Regulation and Control
O Inherited Change
P Selection and Evolution
Q Biodiversity and Conservation
R Gene Technology
S Biotechnology
T Crop Plants
U Aspects of Human Reproduction

The main part of this book, the Content Guidance, summarises the facts and concepts covered by the learning outcomes in all of these 21 sections.

Assessment

The AS examination can be taken at the end of the first year of your course, or with the A2 examination papers at the end of the second year of your course.

What is assessed?

Both the AS and A2 examinations will test three Assessment Objectives. These are:

A: Knowledge with understanding

This involves your knowledge and understanding of the facts and concepts described in the learning outcomes in all of the 21 sections. Questions testing this Assessment Objective will make up 45% of the whole examination.

B: Handling information and solving problems

This requires you to use your knowledge and understanding to answer questions involving unfamiliar contexts or data. The examiners ensure that questions testing this Assessment Objective cannot have been practised by candidates. You will have to *think* to answer these questions, not just remember! An important part of your preparation for the examination will be to gain confidence in answering this kind of question. Questions testing this Assessment Objective will make up 32% of the whole examination.

C: Experimental skills and investigations

This involves your ability to do practical work. The examiners set questions that require you to carry out investigations. It is most important that you take every opportunity to improve your practical skills as you work through your course. Your teacher should give you plenty of opportunity to do practical work in a laboratory. Questions testing this Assessment Objective will make up 23% of the whole examination.

Notice that more than half of the marks in the examination — 55% — are awarded for Assessment Objectives B and C. You need to work hard on developing these skills, as well as learning facts and concepts. There is guidance about Assessment Objective C for AS on pages 98–120, and for A2 on pages 226–236.

The examination papers

The AS examination has three papers:
- Paper 1   Multiple choice
- Paper 2   Structured questions
- Paper 31 or 32   Advanced practical skills

Paper 1 and Paper 2 test Assessment Objectives A and B. Paper 3 tests Assessment Objective C.

Paper 1 contains 40 multiple choice questions. You have 1 hour to answer this paper. This works out at about one question per minute, with time left over to go back through some of the questions again.

Paper 2 contains structured questions. You write your answers on lines provided in the question paper. You have 1 hour 15 minutes to answer this paper.

Paper 31 or 32 is a practical examination. You will work in a laboratory. As for Paper 2, you write your answers on lines provided in the question paper. You have 2 hours to answer this paper.

The A2 examination has two papers:
- Paper 4   Structured questions
- Paper 5   Planning, analysis and evaluation

Paper 4 tests Assessment Objectives A and B. Paper 5 tests Assessment Objective C.

Paper 4 has two sections and you have 2 hours to complete it. Section A contains structured questions. You write your answers on lines provided in the question paper. Section B contains two free-response questions, from which you choose one. Paper 5 is also a written paper. Although it tests your practical skills, you will not be working in a laboratory. You will take this paper in an ordinary examination room. As for Paper 4, you write your answers on lines provided in the question paper. You have 1 hour 15 minutes to answer this paper.

The marks for all the papers, both AS and A2, are given on pages 10 and 11.

You can find copies of past papers at **http://www.cambridgestudents.org.uk/ subjectpages/biology/asalbiology/pastpapers/**

Scientific language

Throughout your biology course, and especially in your examination, it is important to use clear and correct biological language. Scientists take great care to use language precisely. If doctors or researchers do not use exactly the right word when communicating with someone, then what they say could easily be misinterpreted.

Biology has a huge number of specialist terms (probably more than any other subject you can choose to study at AS and A2) and it is important that you learn them and use them.

However, the examiners are testing your knowledge and understanding of biology, not how well you can write in English. They will do their best to understand what you mean, even if some of your spelling and grammar is not correct. Nevertheless, there are some words that you really must spell correctly, because they could be confused with other biological terms. These include:
- words that differ from one another by only one letter, for example amylose and amylase
- words with similar spellings but different meanings, for example glycogen and glucagon; meiosis and mitosis; adenine and adenosine; thymine and thiamine.

In the Syllabus Content section of the syllabus, the words for which you need to know definitions are printed in *italic*. You will find definitions of some of these words in the Definitions section of the syllabus.

Revision

You can download a revision checklist at **http://www.cambridgestudents.org.uk/ subjectpages/biology/asalbiology/** This lists all of the learning outcomes, and you can tick them off or make notes about them as your revision progresses.

There are many different ways of revising, and what works well for you may not be as suitable for someone else. Have a look at the suggestions below and try some of them out.

- **Revise continuously**. Don't think that revision is something you do just before the exam. Life is much easier if you keep revising all through your biology course. Find 15 minutes a day to look back over work you did a few weeks ago, to keep it fresh in your mind. You will find this very helpful when you come to start your intensive revision.

- **Understand it**. Research shows that people learn things much more easily if the brain recognises that they are important and that they make sense. Before you try to learn a topic, make sure that you understand it. If you don't, ask a friend or a teacher, find a different textbook in which to read about it, or look it up on the internet. Work at it until you feel that you understand it and *then* try to learn it.

- **Make your revision active**. Just reading your notes or a textbook will not do any harm, but nor will it do much good. Your brain only puts things into its long-term memory if it thinks they are important, so you need to convince it that they are. You can do this by making your brain *do* something with what you are trying to learn. So, if you are revising from a table comparing eukaryotic and prokaryotic cells, try rewriting it as a paragraph of text, or convert it into a pair of annotated drawings. You will learn much more by constructing your own list of bullet points, flow diagram or table than just trying to remember one that someone else has constructed.

- **Fair shares for all**. Don't always start your revision in the same place. If you always start at the beginning of the course, then you will learn a great deal about cells but not very much about immunity or ecology. Make sure that each part of the syllabus gets its fair share of your attention and time.

- **Plan your time**. You may find it helpful to draw up a revision plan, setting out what you will revise and when. Even if you don't stick to it, it will give you a framework that you can refer to. If you get behind with it, you can rewrite the next parts of the plan to squeeze in the topics you haven't yet covered.

- **Keep your concentration**. It's often said that it is best to revise in short periods, say 20 minutes or half an hour. This is true for many people, if they find it difficult to concentrate for longer than that. But there are others who find it better to settle down for a much longer period of time — even several hours — and really get into their work and stay concentrated without interruptions. Find out which works best for you. It may be different at different times of day. Maybe you can only concentrate well for 30 minutes in the morning, but are able to get lost in your work for several hours in the evening.

- **Don't assume you know it**. The topics where exam candidates are least likely to do well are often the ones that they have already learned something about at GCSE, IGCSE or O level. This is probably because if you think you already know something then you give that a low priority when you are revising. It is important to remember that what you knew for your previous examinations is almost certainly not detailed enough for AS or A2.

The examination

Once you are in the examination room, you can stop worrying about whether or not you have done enough revision. Now you can concentrate on making the best use of the knowledge, understanding and skills that you have built up through your biology course.

Time

On average, you should allow about 1 minute for every mark on the examination paper.

In Paper 1, you will have to answer 40 multiple choice questions in 1 hour. If you work to the 'one-mark-a-minute' rule, that should give you plenty of time to look back over your answers and check any that you weren't quite sure about. Do answer every question, even if you only guess at the answer. Even if you don't know the correct answer, you can probably eliminate one or two of the possible answers, which will increase the chances of your final guess being correct.

In Paper 2, you will have to answer 60 marks-worth of short answer questions in 75 minutes, so once again there should be some time left over to check your answers at the end. It is probably worth spending a short time at the start of the examination to look through the whole paper. If you spot a question that you think may take you a little longer than others (for example, a question that has data to analyse), then you can make sure you allow plenty of time for this one.

In Paper 31 or 32, you will be working in a laboratory. You have 2 hours to answer 40 marks-worth of questions. This is much more time per mark than in the other papers, because you will have to do quite a lot of hands-on practical work before you obtain answers to some of the questions. There will probably be two questions, and you should aim to spend approximately 1 hour on each of them. Your teacher may split the class so that you have to move from one question to the other half way through the time allowed. It's easy to panic in a practical exam, but if you have done plenty of practical work throughout your course this will help you a lot. Do read through the whole question before you start, and do take time to set up your apparatus correctly and to collect your results carefully and methodically.

In Paper 4, you will have to answer 85 marks-worth of structured questions and 15 marks-worth of a free-response question in 120 minutes. This gives you time for the normal one-mark-a-minute, plus 20 minutes spare. You will need these 20 minutes for:

- thinking carefully about unfamiliar data that will be presented in some of the structured questions;
- deciding which of the free-response questions to answer, and planning your answer to it;
- checking through your answers towards the end of the examination.

It is probably worth spending a short time at the start of the examination to look through the whole paper to see if there is a question you think will take longer to answer.

In Paper 5, you will have to answer 30 marks-worth of questions in 75 minutes. This may look like a generous time allowance, as it gives you more than 2 minutes per mark. However, you will be working on two questions that are very likely to contain at least partly unfamiliar material. You will need to spend plenty of time really getting to grips with the questions before you begin to answer them.

Read the question carefully

This sounds obvious, but candidates lose large numbers of marks by not reading the question carefully.

- There is often vital information at the start of the question that you will need in order to answer the questions themselves. Don't just jump straight to the first place where there are answer lines and start writing. Begin by reading from the beginning of the question. Examiners are usually very careful not to give you unnecessary information, so if it is there it is probably needed.
- Do look carefully at the command words at the start of each question, and make sure that you do what they say. For example if you are asked to *explain* something and you only *describe* it, you will not get many marks — indeed, you may not get any marks at all, even if your description is a very good one. You can find the command words and their meanings towards the end of the syllabus.
- Do watch out for any parts of questions that don't have answer lines. For example, you may be asked to label something on a diagram, or to draw a line on a graph, or to write a number in a table. Many candidates miss out these questions and lose significant numbers of marks.

Depth and length of answer

The examiners give you two useful guidelines about how much you need to write.

- **The number of marks** The more marks, the more information you need to give in your answer. If there are 2 marks, you will need to give at least two pieces of correct and relevant information in your answer in order to get full marks. If there are 5 marks, you will need to write much more. But don't just write for the sake of it — make sure that what you write *answers the question*. And don't just keep writing the same thing several times over in different words.

- **The number of lines** This isn't such a useful guideline as the number of marks, but it can still help you to know how much to write. If you find your answer won't fit on the lines, then you probably have not focused sharply enough on the question. The best answers are short, precise, use correct biological language and don't repeat themselves.

Writing, spelling and grammar

The examiners are testing your knowledge and understanding of biology, not your ability to write English. However, if they cannot understand what you have written, they cannot give you any marks. It is your responsibility to communicate clearly. Don't scribble so fast that the examiner cannot read what you have written. Every year, candidates lose marks because the examiner could not read their writing.

Like spelling, grammar is not taken into consideration when marking your answers — so long as the examiner can understand what you are trying to say. One common difficulty is if you use the word 'it' in your answer, and the examiner is not sure what you are referring to. For example, imagine a candidate writes 'A red blood cell contains haemoglobin. It is a protein.' Does the candidate mean that the red blood cell is a protein, or that haemoglobin is a protein? If the examiner cannot be sure, the candidate may not be given the benefit of the doubt.

AS
Content
Guidance

A Cell structure
Microscopy

Light microscopes and electron microscopes

Cells are the basic units from which living organisms are made. Most cells are very small, and their structures can only be seen by using a microscope.

You will use a light microscope during your AS level course. Light rays pass through the specimen on a slide and are focused by an objective lens and an eyepiece lens. This produces a magnified image of the specimen on the retina of your eye. Alternatively, the image can be projected onto a screen, or recorded by a camera.

An electron microscope uses beams of electrons rather than light rays. The specimen has to be very thin and must be placed in a vacuum, to allow electrons to pass through it. The electrons are focused onto a screen, or onto photographic film, where they form a magnified image of the specimen.

Magnification and resolution

Magnification can be defined as:

$$\text{magnification} = \frac{\text{size of image}}{\text{actual size of object}}$$

This can be rearranged to:

$$\text{actual size of object} = \frac{\text{size of image}}{\text{magnification}}$$

There is no limit to the amount you can magnify an image. However, the amount of *useful* magnification depends on the **resolution** of the microscope. This is the ability of the microscope to distinguish two objects as separate from one another. The smaller the objects that can be distinguished, the higher the resolution. Resolution is determined by the wavelength of the rays that are being used to view the specimen. The wavelength of a beam of electrons is much smaller than the wavelength of light. An electron microscope can therefore distinguish between much smaller objects than a light microscope — in other words, an electron microscope has a much higher resolution than a light microscope. We can therefore see much more fine detail of a cell using an electron microscope than using a light microscope.

As cells are very small, we have to use units much smaller than millimetres to measure them. These units are micrometres, **μm**, and nanometres, **nm**.

$1 \text{ mm} = 1 \times 10^{-3} \text{ m}$

$1 \text{ μm} = 1 \times 10^{-6} \text{ m}$

$1 \text{ nm} = 1 \times 10^{-9} \text{ m}$

To change mm into μm, multiply by 1000.

Magnification calculations

You should be able to work out the real size of an object if you are told how much it has been magnified.

For example, this drawing of a mitochondrion has been magnified 100 000 times.

- Use your ruler to measure its length in mm. It is 50 mm long.
- As it is a very small object, convert this measurement to µm by multiplying by 1000.

$$50 \times 1000 = 50\,000 \, \mu m.$$

- Substitute into the equation:

$$\text{actual size of object} = \frac{\text{size of image}}{\text{magnification}}$$

$$= \frac{50\,000}{100\,000}$$

$$= 0.5 \, \mu m$$

You can also use a scale bar to do a similar calculation for this drawing of a chloroplast.

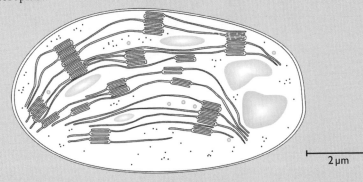

2 µm

- Measure the length of the scale bar.
- Calculate its magnification using the formula

$$\text{magnification} = \frac{\text{size of image}}{\text{actual size of object}}$$

$$= \frac{\text{length of scale bar}}{\text{length the scale bar represents}}$$

$$= \frac{200\,000}{2}$$

$$= \times 100\,000$$

- Measure the length of the image of the chloroplast in mm, and convert to µm. You should find that it is 80 000 µm long.
- Calculate its real length using the formula

$$\text{actual size of object} = \frac{\text{size of image}}{\text{magnification}}$$

$$= \frac{80\,000}{10\,000}$$

$$= 8\,\mu m$$

Measuring cells using a graticule

An eyepiece graticule is a little scale bar that you can place in the eyepiece of your light microscope. When you look down the microscope, you can see the graticule as well as the specimen.

The graticule is marked off in 'graticule units', so you can use the graticule to measure the specimen you are viewing in these graticule units. Just turn the eyepiece so that the graticule scale lies over the object you want to measure. It will look like this:

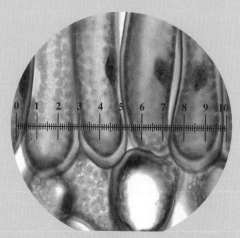

We can say that the width of one cell is 23 graticule units.

The graticule units have to be converted to real units, such as mm or µm. This is called **calibration**. To do this, you use a special slide called a **stage micrometer** that is marked off in a tiny scale. There should be information on the slide that tells you the units in which it has been marked. The smallest markings are often 0.01 mm apart — that is, 10 µm apart.

Take the specimen off the stage or the microscope and replace it with the stage micrometer. Focus on it using the same objective lens as you used for viewing the specimen.

Line up the micrometer scale and the eyepiece graticule scale. You can do this by turning the eyepiece, and by moving the micrometer on the stage. Make sure that two large markings on each scale are exactly lined up with each other. You should be able to see something like this:

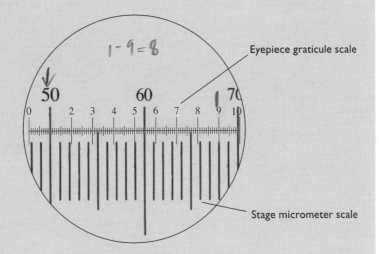

- You can see that the 50 mark on the stage micrometer is lined up with the 1.0 mark on the eyepiece graticule. Work along towards the right until you see another two lines that are exactly lined up. There is a good alignment of 68 on the stage micrometer and 9.0 on the eyepiece graticule. So you can say that:

80 small eyepiece graticule markings = 18 stage micrometer markings

$$= 18 \times 0.01\,\text{mm} = 0.18\,\text{mm}$$

$$= 180\,\mu\text{m}$$

so 1 small eyepiece graticule marking = $\frac{180}{80}$ = 2.25 μm.

Now we can calculate the real width of the plant cell we measured. It was 23 eyepiece graticule units long. So its real width is:

$$23 \times 2.25 = 51.75\,\mu\text{m}$$

If you want to look at something using a different objective lens, you will have to do the calibration of eyepiece graticule units all over again using this lens. Once you have done it, you can save your calibrations for the next time you use the same microscope with the same eyepiece graticule and the same objective lens.

Cell structure and function

The diagrams show a typical animal cell and a typical plant cell as seen using an electron microscope.

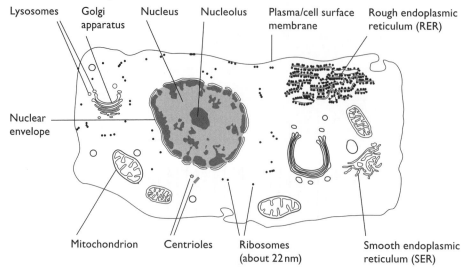

A typical animal cell, ×2000

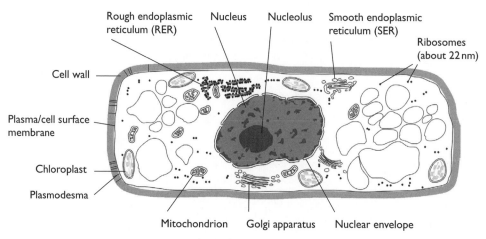

A typical plant cell, ×1500

Functions of membrane systems and organelles

The **plasma membrane**, also known as the **cell surface membrane**, controls what enters and leaves the cell. Its structure and functions are described in detail on pages 45–51. There are also many membranes within the cell, which help to make different compartments in which different chemical reactions can take place without interfering with one another.

The **nucleus** is surrounded by a pair of membranes, which make up the **nuclear envelope**. The nucleus contains **chromosomes**, each of which contains a very long molecule of DNA. The DNA determines the sequences in which amino acids are linked together in the cytoplasm to form protein molecules. This is described on pages 57–61.

Within the nucleus there is a darker area (not surrounded by its own membrane) called the **nucleolus**. This is where new ribosomes are made, following a code on part of the DNA.

Ribosomes are small structures made of RNA and protein. They are found free in the cytoplasm, and also attached to **rough endoplasmic reticulum** (**RER**). The RER is an extensive network of membranes in the cytoplasm. The membranes enclose small spaces called **cisternae**. Proteins are made on the ribosomes, by linking together amino acids.

If the proteins are to be processed or exported from the cell, the growing chains of amino acids move into the cisternae of the RER as they are made. The cisternae then break off to form little **vesicles** that travel to the **Golgi apparatus**. Here they may be modified, for example by adding carbohydrate groups to them. Vesicles containing the modified proteins break away from the Golgi apparatus and are transported to the cell surface membrane, where they are secreted from the cell by exocytosis (pages 50–51).

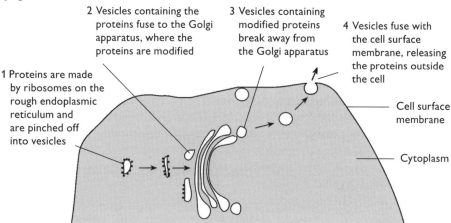

The interrelationship between RER and the Golgi apparatus

Smooth endoplasmic reticulum (**SER**) is usually less extensive than RER. It does not have ribosomes attached to it, and the cisternae are usually more flattened than those of the RER. It is involved in the synthesis of steroid hormones and the breakdown of toxins.

Mitochondria have an envelope (two membranes) surrounding them. The inner one is folded to form **cristae**. This is where aerobic respiration takes place, producing ATP. The first stage of this process, called the Krebs cycle, takes place in the **matrix**.

The final stage, oxidative phosphorylation, takes place on the membranes of the cristae.

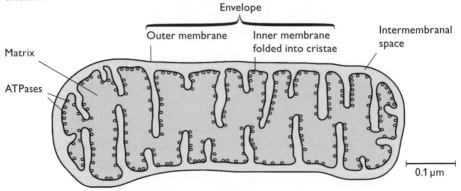

Longitudinal section through a mitochondrion

Lysosomes are little membrane-bound packages of hydrolytic (digestive) enzymes. They form by breaking off from the Golgi apparatus. They are used to digest bacteria or other cells taken into the cell by phagocytosis, or to break down unwanted or damaged organelles within the cell.

Centrioles are found only in animal cells, not plant cells. They are made of tiny tubules called microtubules, arranged in a circular pattern. The two centrioles lie at right angles to one another. It is from here that the microtubules are made that form the spindle during cell division in animal cells.

Chloroplasts are found only in some plant cells. Like mitochondria, they are surrounded by an envelope made up of two membranes. Their background material is called the **stroma**, and it contains many paired membranes called **thylakoids**. In places, these form stacks called **grana**. The grana contain chlorophyll, which absorbs energy from sunlight. The first reactions in photosynthesis, called the light-dependent reactions and photophosphorylation, take place on the membranes. The final stages, called the Calvin cycle, take place in the stroma. Chloroplasts often contain starch grains, which are storage materials formed from the sugars that are produced in photosynthesis.

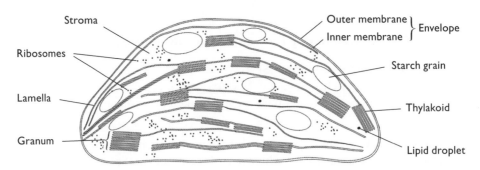

Longitudinal section through a chloroplast

Plant cells are always surrounded by a **cell wall** made of cellulose, which is never found around animal cells. The structure of cellulose is described on page 26.

Drawing plan diagrams

An organ usually contains many different types of cells. These are arranged in a particular pattern characteristic of the organ, with cells of a similar type found together, forming distinctive tissues.

A plan diagram shows the outline of the various tissues in an organ such as a leaf or an eye. It does **not** show individual cells.

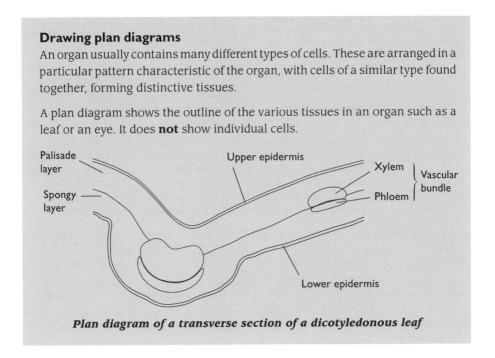

Plan diagram of a transverse section of a dicotyledonous leaf

Prokaryotic cells

Prokaryotic cells are found in bacteria and archaea, whereas eukaryotic cells are found in animals, plants, protista and fungi. Prokaryotic cells are generally much smaller than eukaryotic cells. The fundamental difference between prokaryotic and eukaryotic cells is that they do not have a nucleus or any other membrane-bound organelles.

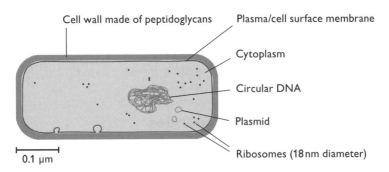

Structure of a prokaryotic cell

Comparison of prokaryotic, animal and plant cells

Feature	Prokaryotic cells	Eukaryotic cells	
		Animal cells	**Plant cells**
Plasma/cell surface membrane	Always present	Always present	Always present
Cell wall	Always present; made up of peptidoglycans	Never present	Always present; made up of cellulose
Nucleus and nuclear envelope	Never present	Always present	Always present
Chromosomes	Contain so-called 'bacterial chromosomes' — a circular molecule of DNA not associated with histones; bacteria may also contain smaller circles of DNA called plasmids	Contain several chromosomes, each made up of a linear DNA molecule associated with histones	Contain several chromosomes, each made up of a linear DNA molecule associated with histones
Mitochondria	Never present	Usually present	Usually present
Chloroplasts	Never present, though some do contain chlorophyll or other photosynthetic pigments	Never present	Sometimes present
Rough and smooth endoplasmic reticulum and Golgi apparatus	Never present	Usually present	Usually present
Ribosomes	Present, about 18 nm diameter	Present, about 22 nm diameter	Present, about 22 nm diameter
Centrioles	Never present	Usually present	Never present

B Biological molecules
Carbohydrates

Carbohydrates are substances whose molecules contain carbon, hydrogen and oxygen atoms, and in which there are approximately twice as many hydrogen atoms as carbon or oxygen atoms.

Monosaccharides and disaccharides

The simplest carbohydrates are **monosaccharides**. These are sugars. They include glucose, fructose and galactose. These three monosaccharides each have six carbon atoms, so they are also known as hexose sugars. Their molecular formula is $C_6H_{12}O_6$.

Monosaccharide molecules are often in the form of a ring made up of carbon atoms and one oxygen atom. Glucose molecules can take up two different forms, called α-glucose and β-glucose.

Structural formulae of α-glucose and β-glucose molecules

Two monosaccharides can link together to form a **disaccharide**. For example, two glucose molecules can link to produce **maltose**. The bond that joins them together is called a **glycosidic bond**. As the two monosaccharides react and the glycosidic bond forms, a molecule of water is released. This type of reaction is known as a **condensation reaction**. Different disaccharides can be formed by linking different monosaccharides.

Disaccharide	Monosaccharides
Maltose	Glucose + Glucose
Lactose	Glucose + Galactose
Sucrose	Glucose + Fructose

Formation of maltose by a condensation reaction

Formation of sucrose by a condensation reaction

Disaccharides can be split apart into two monosaccharides by breaking the glyco-sidic bond. To do this, a molecule of water is added. This is called an **hydrolysis reaction**.

Breakdown of a disaccharide by an hydrolysis reaction

Functions of monosaccharides and disaccharides
- Monosaccharides and disaccharides are good sources of energy in living organ-isms. They can be used in respiration, in which the energy they contain is used to make ATP.
- Because they are soluble, they are the form in which carbohydrates are trans-ported through an organism's body. In animals, glucose is transported dissolved in blood plasma. In plants, sucrose is transported in phloem sap.

All monosaccharides and some disaccharides act as reducing agents, and will reduce blue Benedict's solution to produce an orange-red precipitate. They are called **reducing sugars**. Sucrose is a **non-reducing sugar**.

Polysaccharides

These are substances whose molecules contain hundreds or thousands of monosac-charides linked together into long chains. Because their molecules are so enormous, the majority do not dissolve in water. This makes them good for storing energy (starch and glycogen) or for forming strong structures (cellulose).

AS content guidance

Storage polysaccharides

In animals and fungi, the storage polysaccharide is **glycogen**. It is made of α-glucose molecules linked together by glycosidic bonds. Most of the glycosidic bonds are between carbon 1 on one glucose, and carbon 4 on the next, so they are called 1–4 links. There are also some 1–6 links, which form branches in the chain. When needed, the glycosidic bonds can be hydrolysed by carbohydrase enzymes to form monosaccharides, which can be used in respiration. The branches mean there are many 'ends', which increases the rate at which carbohydrases can hydrolyse the molecules.

1–4 link   1–6 link

A small part of a glycogen molecule

In plants, the storage polysaccharide is **starch**. Starch is a mixture of two substances, **amylose** and **amylopectin**. An amylose molecule is a very long chain of α-glucose molecules with 1–4 links. It coils up into a spiral, making it very compact. The spiral is held in shape by hydrogen bonds between small charges on some of the hydrogen and oxygen atoms in the glucose units. An amylopectin molecule is very similar to glycogen.

amylose = a-glucose with 1-4 links

Amylose contains chains of α 1–4 linked glucose

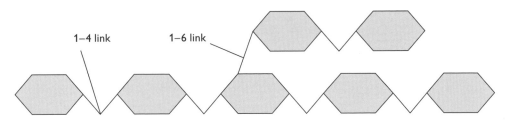

The glucose molecules in a chain all have the same orientation

α 1–4 glycosidic bond

amylopectin — contains 1,4 + 1,6 links like glycogen

1–4 link

Hydrogen bond — many of these hold the amylose in a spiral shape

A small part of an amylose molecule

Structural polysaccharides

Plant cell walls contain the polysaccharide **cellulose**. Like amylose, this is made of many glucose molecules linked by glycosidic bonds between carbon 1 and carbon 4. However, in cellulose the glucose molecules are in the β form. This means that adjacent glucose molecules in the chain are upside-down in relation to one another. The chain stays straight, rather than spiralling. Hydrogen bonds form between different chains. This causes the chains to associate into bundles called **microfibrils**.

β 1–4 glycosidic bond

$^6CH_2OH$ H   OH $^6CH_2OH$

The glucose molecules in a chain alternate in their orientation (look at the numbered carbon atoms)

Cellulose contains chains of β 1–4 linked glucose

Structurally strong

Parallel chains of β 1–4 linked glucose

Hydrogen bonds hold the chains together to form microfibrils

Cellulose molecules

The resulting microfibrils are very strong. This makes cellulose an excellent material for plant cell walls, because it will not break easily if the plant cell swells as it absorbs water. The microfibrils are also very difficult to digest, because few organisms have an enzyme that can break the β 1–4 glycosidic bonds.

Tests for carbohydrates

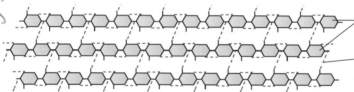

(structurally strong + difficult to digest)

Reducing sugar

Add Benedict's reagent and heat. An orange-red precipitate indicates the presence of reducing sugar. If standard volumes of the unknown solutions and excess Benedict's reagent are used, the mass of precipitate or intensity of the orange-red colour indicates the concentration of the solution. This can be matched against colour standards, prepared using reducing sugar solutions of known concentration.

Non-reducing sugar

This test should only be done on solutions known not to contain reducing sugars. Hydrolyse by heating with dilute HCl, then neutralise with sodium hydrogen-carbonate. Then carry out the test for reducing sugar. then add benedicts reagentt neat

Starch

Add iodine in potassium iodide solution. A blue-black colour indicates the presence of starch.

Lipids

Lipids, like carbohydrates, also contain carbon, hydrogen and oxygen, but there is a much smaller proportion of oxygen. Lipids include triglycerides and phospholipids. All lipids are insoluble in water.

Triglycerides

A triglyceride molecule is made of a 'backbone' of glycerol, to which three fatty acids are attached by ester bonds.

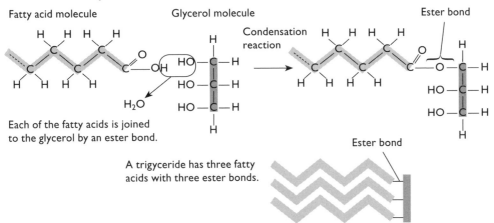

The formation of a triglyceride molecule

Fatty acids have long chains made of carbon and hydrogen atoms. Each carbon atom has four bonds. Usually, two of these bonds are attached to other carbon atoms, and the other two to hydrogen atoms. In some cases, however, there may be only one hydrogen atom attached. This leaves the carbon atom with a 'spare' bond, which attaches to the next-door carbon atom (which also has one less hydrogen bonded to it) forming a double bond. Fatty acids with one or more carbon-carbon double bonds are called unsaturated fatty acids, because they do not contain quite as much hydrogen as they could. Fatty acids with no double bonds are called saturated fatty acids.

Unsaturated and saturated fatty acids

Lipids containing unsaturated fatty acids are called unsaturated lipids, and those containing completely saturated fatty acids are called saturated lipids. Animal lipids

animal lipids ≐ saturated
(C−C)

are often saturated lipids. Plant lipids are often unsaturated. Unsaturated lipids tend to have lower melting points than saturated lipids.

Triglycerides are used as energy storage compounds in plants, animals and fungi. Their insolubility in water helps to make them suitable for this function. They contain more energy per gram than polysaccharides, so can store more energy in less mass. In mammals, stores of triglycerides often build up beneath the skin, in the form of adipose tissue. The cells in adipose tissue contain oil droplets made up of triglycerides.

This tissue also helps to insulate the body against heat loss. It is a relatively low-density tissue, and therefore increases buoyancy. These properties make it especially useful for aquatic mammals that live in cold water, such as whales and seals.

Adipose tissue also forms a protective layer around some of the body organs, for example the kidneys.

In plants, triglycerides often make up a major part of the energy stores in seeds, either in the cotyledons (e.g. in sunflower seeds) or in the endosperm (e.g. in castor beans).

Phospholipids

A phospholipid molecule is like a triglyceride in which one of the fatty acids is replaced by a phosphate group.

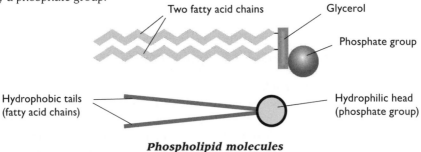

Phospholipid molecules

The fatty acid chains have no electrical charge and so are not attracted to the dipoles of water molecules (see page 45). They are said to be **hydrophobic**.

The phosphate group has an electrical charge and is attracted to water molecules. It is **hydrophilic**.

In water, a group of phospholipid molecules therefore arranges itself into a bilayer, with the hydrophilic heads facing outwards into the water and the hydrophobic tails facing inwards, therefore avoiding contact with water.

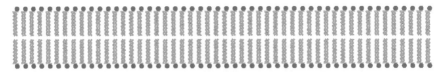

A phospholipid bilayer

This is the basic structure of a cell membrane. The functions of phospholipids in membranes are described on pages 45-51.

Test for lipids

Mix the substance to be tested with absolute ethanol. Decant the ethanol into water. A milky emulsion indicates the presence of lipid.

Proteins

Proteins are large molecules made of long chains of amino acids.

Amino acids

All amino acids have the same basic structure, with an amine group and a carboxyl group attached to a central carbon atom. There are twenty different types of amino acid, which differ in the atoms present in the R group. In the simplest amino acid, glycine, the R group is a single hydrogen atom.

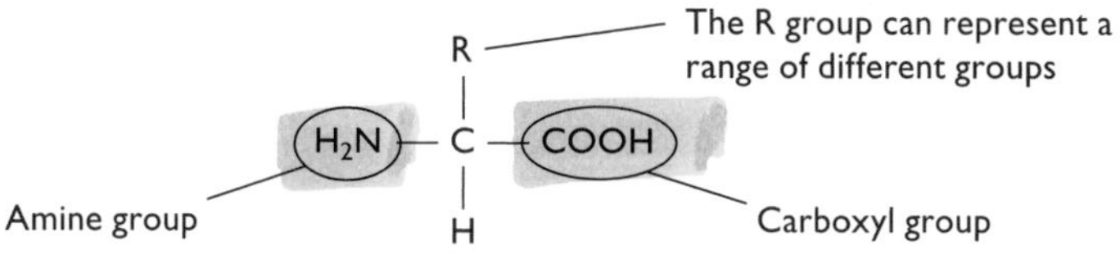

An amino acid

Two amino acids can link together by a condensation reaction to form a dipeptide. The bond that links them is called a **peptide bond**, and water is produced in the reaction.

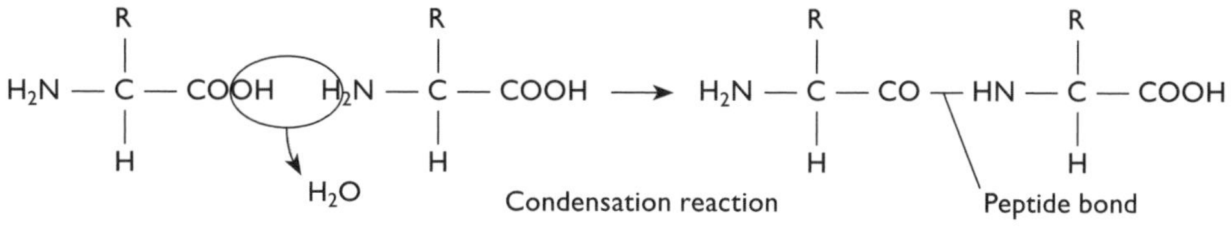

Formation of a dipeptide

The dipeptide can be broken down in a hydrolysis reaction, which breaks the peptide bond with the addition of a molecule of water.

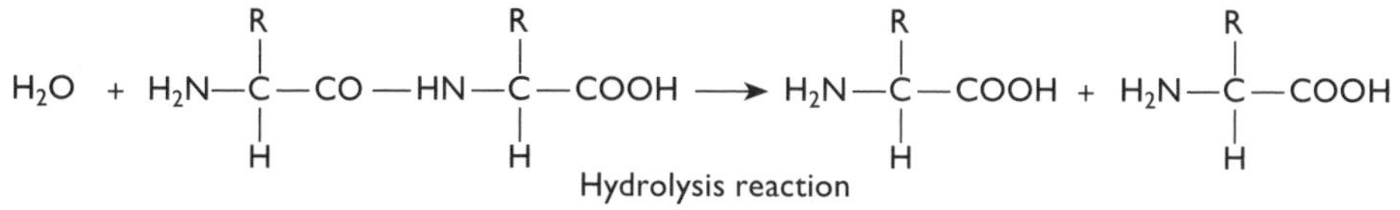

Breakdown of a dipeptide

Structure of protein molecules

Amino acids can be linked together in any order to form a long chain called a **polypeptide**. A polypeptide may form a protein molecule on its own, or it may associate with other polypeptides to form a protein molecule.

The sequence of amino acids in a polypeptide or protein molecule is called its **primary structure**. Note — the three letters in each box are the first three letters of the amino acid, e.g. Val is valine, Leu is leucine.

Val – Leu – Ser – Pro – Ala – Asp – Lys – Thr – Asn – Val – Lys – Ala

The primary structure of a small part of a polypeptide

The chain of amino acids often folds or curls up on itself. For example, many polypeptide chains coil into a regular 3D shape called an **alpha helix**. This is held in shape by **hydrogen bonds** between amino acids at different places in the chain. This regular shape is an example of **secondary structure** of a protein. Another example is the beta-pleated strand.

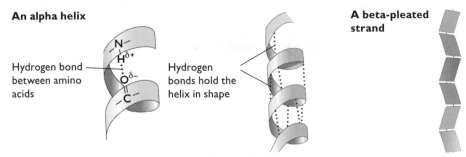

An alpha helix

Hydrogen bond between amino acids

Hydrogen bonds hold the helix in shape

A beta-pleated strand

Examples of secondary structure

The polypeptide chain may also fold around on itself to form a more complex three-dimensional shape. This is called the **tertiary structure** of the protein. Once again, hydrogen bonds between amino acids at different points in the chain help to hold it in its particular 3-D shape. There are also other bonds involved, including **ionic bonds, disulfide bonds** and **hydrophobic interactions**.

The globular shape of this polypeptide is an example of tertiary structure

Tertiary structure of a protein

Many proteins are made of more than one polypeptide chain. These chains are held together by the same type of bonds as in the tertiary structure. The overall structure of the molecule is known as the **quaternary structure** of the protein. The tertiary and quaternary structures of a protein, and therefore its properties, are ultimately determined by its primary structure.

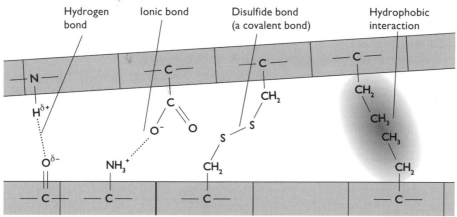

Bonds involved in maintaining the secondary, tertiary and quaternary structure of proteins

Globular and fibrous proteins

Metabolic   Structural

soluble

Globular proteins have molecules that fold into a roughly spherical three-dimensional shape. Examples include haemoglobin, insulin and enzymes. They are often soluble in water and may be physiologically active — that is, they are involved in metabolic reactions within or outside cells.

Fibrous proteins have molecules that do not curl up into a ball. They have long, thin molecules, which often lie side by side to form fibres. Examples include keratin (in hair) and collagen (in skin and bone). They are not soluble in water and are not generally physiologically active. They often have structural roles.

insoluble

Haemoglobin — a globular protein

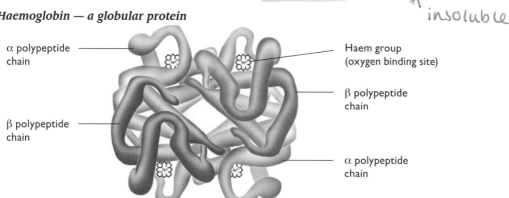

α polypeptide chain

Haem group (oxygen binding site)

β polypeptide chain

β polypeptide chain

α polypeptide chain

The structure of haemoglobin

Relationship between structure and function in haemoglobin

The function of haemoglobin is the transport of oxygen from the lungs to respiring tissues. It is found inside red blood cells.

tertiary is responsible

- **Solubility** The tertiary structure of haemoglobin makes it soluble. The four polypeptide chains are coiled up so that R groups with small charges on them (hydrophilic groups) are on the outside of the molecule. They therefore form hydrogen bonds with water molecules. Hydrophobic R groups are mostly found inside the molecule.

haem group responsible

- **Ability to combine with oxygen** The haem group contained within each polypeptide chain enables the haemoglobin molecule to combine with oxygen. Oxygen molecules combine with the iron ion, Fe^{2+}, in the haem group. One oxygen molecule (two oxygen atoms) can combine with each haem group, so one haemoglobin molecule can combine with four oxygen molecules (eight oxygen atoms).

- **Pick-up and release of oxygen** The overall shape of the haemoglobin molecule enables it to pick up oxygen when the oxygen concentration is high, and to release oxygen when the oxygen concentration is low. Small changes in oxygen concentration have a large effect on how much oxygen the haemoglobin molecule can hold. Once one oxygen molecule has combined with one haem group, the whole molecule changes its shape in such a way that it is easier for oxygen to combine with the other three haem groups. (See also information about the oxygen dissociation curve for haemoglobin on pages 72–73.)

Collagen — a fibrous protein

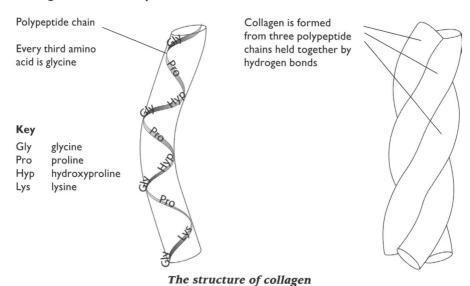

Polypeptide chain

Every third amino acid is glycine

Collagen is formed from three polypeptide chains held together by hydrogen bonds

Key

Gly	glycine
Pro	proline
Hyp	hydroxyproline
Lys	lysine

The structure of collagen

Relationship between structure and function in collagen

The function of collagen is to provide support and some elasticity in many different animal tissues, such as human skin, bone and tendons.

- **Insolubility** Collagen molecules are very long and are too large to be able to dissolve in water.

- **High tensile strength** Three polypeptide chains wind around one another, held together by hydrogen bonds, to form a three-stranded molecule that can withstand quite high pulling forces without breaking. This structure also allows the molecules to stretch slightly when pulled.
- **Compactness** Every third amino acid in each polypeptide is glycine, whose R group is just a single hydrogen molecule. Their small size allows the three polypeptide chains in a molecule to pack very tightly together.
- **Formation of fibres** There are many lysine molecules in each polypeptide, facing outwards from the three-stranded molecule. This allows covalent bonds to form between the lysine R groups of different collagen molecules, causing them to associate to form fibres.

Test for proteins

Add biuret solution. A purple colour indicates the presence of protein.

PURPLE = PROTEIN PRESENT

Water

About 80% of the body of an organism is water. Water has unusual properties compared with other substances, because of the structure of its molecules. Each water molecule has a small negative charge (δ^-) on the oxygen atom and a small positive charge (δ^+) on each of the hydrogen atoms. This is called a **dipole**.

There is an attraction between the δ^- and δ^+ parts of neighbouring water molecules. This is called a **hydrogen bond**.

A single water molecule

Hydrogen bonding between water molecules

Hydrogen bond

Water molecules

Solvent properties of water

Dipoles = Solvent

The dipoles on water molecules make water an excellent solvent. For example, if you stir sodium chloride into water, the sodium and chloride ions separate and spread between the water molecules — they dissolve in the water (see diagrams on next page). This happens because the positive charge on each sodium ion is attracted to the small negative charge on the oxygen of the water molecules. Similarly, the negative chloride ions are attracted to the small positive charge on the hydrogens of the water molecules.

Attraction between different ions

Any substance that has fairly small molecules with charges on them, or that can separate into ions, can dissolve in water.

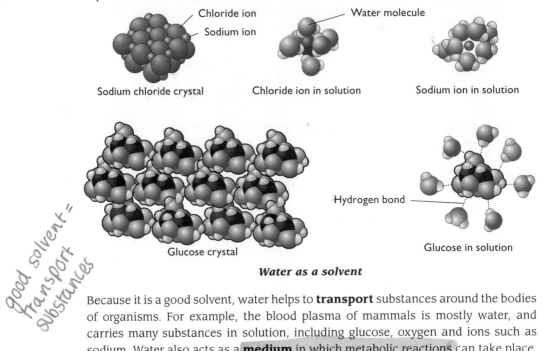

Water as a solvent

Because it is a good solvent, water helps to **transport** substances around the bodies of organisms. For example, the blood plasma of mammals is mostly water, and carries many substances in solution, including glucose, oxygen and ions such as sodium. Water also acts as a **medium** in which metabolic reactions can take place, as the reactants are able to dissolve in it.

Thermal properties of water

- **Water is liquid at normal Earth temperatures**. The hydrogen bonds between water molecules prevent them flying apart from each other at normal temperatures on Earth. Between 0 °C and 100 °C, water is in the liquid state. The water molecules move randomly, forming transitory hydrogen bonds with each other. Other substances whose molecules have a similar structure, such as hydrogen sulfide (H_2S) are gases at these temperatures, because there are no hydrogen bonds to attract their molecules to each other.

- **Water has a high latent heat of evaporation**. When a liquid is heated, its molecules gain kinetic energy, moving faster. Those molecules with the most energy are able to escape from the surface and fly off into the air. A great deal of heat energy has to be added to water molecules before they can do this, because the hydrogen bonds between them have to be broken. When water evaporates, it therefore absorbs a lot of heat from its surroundings. The evaporation of water from the skin of mammals when they sweat therefore has a cooling effect. Transpiration from plant leaves is important in keeping them cool in hot climates.

- **Water has a high specific heat capacity**. Specific heat capacity is the amount of heat energy that has to be added to a given mass of a substance to raise its temperature by 1 °C. Temperature is related to the kinetic energy of the molecules

— the higher their kinetic energy, the higher the temperature. A lot of heat energy has to be added to water to raise its temperature, because much of the heat energy is used to break the hydrogen bonds between water molecules, not just to increase their speed of movement. This means that bodies of water, such as oceans or a lake, do not change their temperature as easily as air does. It also means that the bodies of organisms, which contain large amounts of water, do not change temperature easily.

- **Water freezes from the top down.** Like most substances, liquid water becomes more dense as it cools, because the molecules lose kinetic energy and get closer together. However, when it becomes a solid (freezes), water becomes less dense than it was at 4 °C, because the molecules form a lattice in which they are more widely spaced than in liquid water at 4 °C. Ice therefore floats on water. The layer of ice then acts as an insulator, slowing down the loss of heat from the water beneath it, which tends to remain at 4 °C. The water under the ice therefore remains liquid, allowing organisms to continue to live in it even when air temperatures are below the freezing point of water.

Inorganic ions

Inorganic ions in living organisms

Inorganic ion	Examples of functions
Calcium	• as calcium phosphate, provides a hard, strong insoluble matrix in bones and teeth in mammals • required for blood clotting in mammals • as calcium pectate in plant cell walls, forms a matrix in which cellulose fibres lie • involved in the transmission of action potentials from one neurone to another, and in muscle contraction
Sodium	• constantly pumped out of cells by active transport in exchange for potassium ions, providing a positive charge outside the cell which is important in the transmission of nerve impulses
Potassium	• constantly pumped into cells by active transport in exchange for sodium ions; important in the transmission of nerve impulses
Magnesium	• forms part of the chlorophyll molecule in plants, important for the absorption of light energy to drive the reactions of photosynthesis
Chloride	• moved out of cells lining the lungs and digestive system to provide a low water potential outside the cell, causing water to follow so that mucus is not too thick and stiff (failure of this mechanism is the cause of cystic fibrosis)
Nitrate	• used to produce amino acids (and therefore proteins) from the carbohydrates made in plants by photosynthesis
Phosphate	• the production of nucleic acids (DNA and RNA) in cells • the production of ATP (the energy currency of all cells) • the production of phospholipids, essential in cell membranes

C Enzymes

An enzyme is a protein that acts as a biological catalyst — that is, it speeds up a metabolic reaction without itself being permanently changed.

The substance present at the start of an enzyme-catalysed reaction is called the **substrate**. The **product** is the new substance or substances formed.

Active sites

Enzymes are globular proteins. In one part of the molecule, there is an area called the **active site**, where the substrate molecule can bind. This produces an enzyme-substrate complex. The 3-D shape of the active site fits the substrate perfectly, so only one type of substrate can bind with the enzyme. The enzyme is therefore **specific** for that substrate.

The R groups of the amino acids at the active site are able to form temporary bonds with the substrate molecule. This pulls the substrate molecule slightly out of shape, causing it to react and form products.

Activation energy

Substrates generally need to be supplied with energy to cause them to change into products. The energy required to do this is called **activation energy**. In a laboratory, you might supply energy by heating to cause two substances to react together.

Enzymes are able to make substances react even at low temperatures. They reduce the activation energy needed to make the reaction take place. They do this by distorting the shape of the substrate molecule when it binds at the enzyme's active site.

Following the course of an enzyme-catalysed reaction

You can follow what happens over time in a reaction catalysed by an enzyme by:

- measuring the rate of formation of the product
- measuring the rate of disappearance of the substrate.

For example, you can measure the rate of formation of oxygen in this reaction:

$$\text{hydrogen peroxide} \xrightarrow{\text{catalase}} \text{oxygen} + \text{water}$$

All biological material contains catalase. You could mash up some potato tuber or celery stalks, mix them with water and filter the mixture to obtain a solution containing catalase. This can then be added to hydrogen peroxide in a test tube. Use relatively small tubes, so that there is not too much gas in the tube above the liquid.

You could measure the rate of oxygen formation by collecting the gas in a gas syringe and recording the volume every minute until the reaction stops. Note — don't worry if you don't have gas syringes. You could collect the oxygen in an inverted measuring cylinder over water instead.

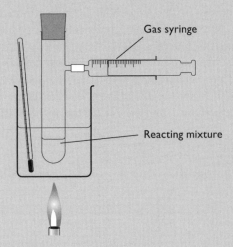

Gas syringe

Reacting mixture

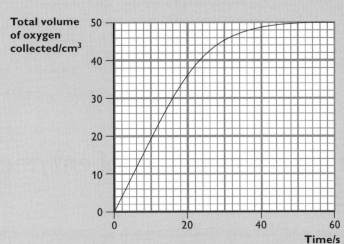

Following the time course of the breakdown of hydrogen peroxide by catalase

You could measure the rate of disappearance of starch in the reaction:

$$\text{starch} \xrightarrow{\text{amylase}} \text{maltose}$$

Add amylase solution to starch suspension in a test-tube. Take samples of the reacting mixture at regular time intervals, and test for the presence of starch using iodine in potassium iodide solution. If starch is still present, you will obtain a blue-black colour. If there is no starch present, the iodine solution will remain orange-brown.

To obtain quantitative results, you could use a colorimeter. Put some of the iodine solution into one of the colorimeter tubes, place it in the colorimeter and adjust the dial to give a reading of 0. This is your standard, with no starch. Every minute, take a sample of the liquid from the starch-amylase mixture and add it to a clean colorimeter tube containing iodine solution. Mix thoroughly, then measure the absorbance. The darker the blue-black colour, the greater the absorbance, and the greater the concentration of starch.

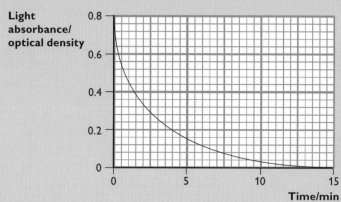

Following the time course of the breakdown of starch by amylase

Factors affecting the rate of enzyme-catalysed reactions

When an enzyme solution is added to a solution of its substrate, the random movements of enzyme and substrate molecules cause them to collide with each other.

As time passes, the quantity of substrate decreases, because it is being changed into product. This decrease in substrate concentration means that the frequency of collisions between enzyme and substrate molecules decreases, so the rate of the reaction gradually slows down. The reaction rate is fastest right at the start of the reaction, when substrate concentration is greatest.

When comparing reaction rates of an enzyme in different circumstances, we should therefore try to measure the initial rate of reaction — that is, the rate of reaction close to the start of the reaction.

Temperature

At low temperatures, enzyme and substrate molecules have little kinetic energy. They move slowly, and so collide infrequently. This means that the rate of reaction is low. If the temperature is increased, then the kinetic energy of the molecules increases. Collision frequency therefore increases, causing an increase in the rate of reaction.

Above a certain temperature, however, hydrogen bonds holding the enzyme molecule in shape begin to break. This causes the tertiary structure of the enzyme to change, an effect called **denaturation**. This affects the shape of its active site. It becomes less likely that the substrate molecule will be able to bind with the enzyme, and the rate of reaction slows down.

The temperature at which an enzyme works most rapidly, just below that at which denaturation begins, is called its optimum temperature. Enzymes in the human body generally have an optimum temperature of about 37 °C, but enzymes from organisms that have evolved to live in much higher or lower temperatures may have much higher or lower optimum temperatures.

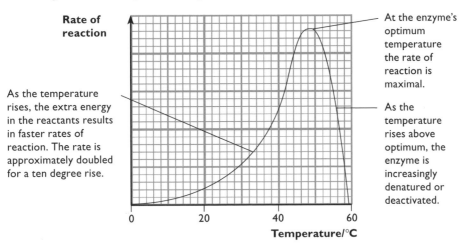

How temperature affects the rate of an enzyme-catalysed reaction

Investigating the effect of temperature on enzyme activity

You can use almost any enzyme reaction for this, such as the action of catalase on hydrogen peroxide as described on pages 35–36. You could use the same method of collecting the gas that is described there, but here is another possible method:

Set up several small conical flasks containing the same volume of hydrogen peroxide solution of the same concentration. Stand each one in a water bath at a particular temperature. Use at least five different temperatures over a good range — say between 0 °C and 90 °C. (If time allows, set up three sets of tubes at each temperature. You will then be able to calculate the mean result for each temperature, which will give you a more representative finding.)

Take a set of test tubes and add the same volume of catalase solution to each one. Stand these in the same set of water baths.

Leave all the flasks and tubes to come to the correct temperature. Check with a thermometer.

Take the first flask, dry its base and sides and stand it on a sensitive top-pan balance. Pour in the solution containing catalase (see page 36) that is at the same temperature, and immediately take the balance reading. Record the new balance readings every 30 seconds (or even more frequently if you can manage it) for about 3 minutes. The readings will go down as oxygen is given off.

Repeat with the solutions kept at each of the other temperatures.

Work out the initial rate of each reaction, either taken directly from your readings, or by drawing a graph of mass lost (which is the mass of oxygen) against time for each temperature, and then working out the gradient of the graph over the first 30 seconds or 60 seconds of the reaction.

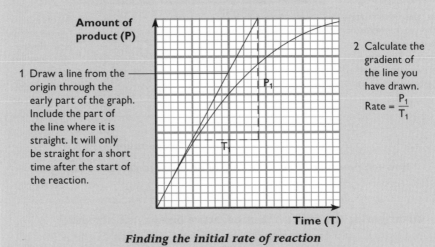

Amount of product (P)

1 Draw a line from the origin through the early part of the graph. Include the part of the line where it is straight. It will only be straight for a short time after the start of the reaction.

2 Calculate the gradient of the line you have drawn.

$$\text{Rate} = \frac{P_1}{T_1}$$

Time (T)

Finding the initial rate of reaction

Now you can use your results to plot a graph of initial rate of reaction (y-axis) against temperature.

pH

pH affects ionic bonds that hold protein molecules in shape. Because enzymes are proteins, their molecules are affected by changes in pH. Most enzyme molecules only maintain their correct tertiary structure within a very narrow pH range, generally around pH 7. Some, however, require a very different pH; one example is the protein-digesting enzyme pepsin found in the human stomach, which has an optimum pH of 2.

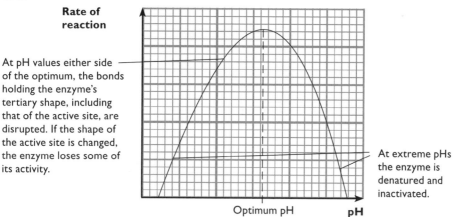

At pH values either side of the optimum, the bonds holding the enzyme's tertiary shape, including that of the active site, are disrupted. If the shape of the active site is changed, the enzyme loses some of its activity.

At extreme pHs the enzyme is denatured and inactivated.

How pH affects the rate of an enzyme-catalysed reaction

Investigating the effect of pH on enzyme activity

You can adapt the method described on pages 39–40 for investigating the effect of temperature on the rate of breakdown of hydrogen peroxide by catalase.

Vary pH by using different buffer solutions added to each enzyme solution. Keep temperature, enzyme concentration, substrate concentration and total volume of reactants the same for all the tubes. Record, process and display results as before.

Enzyme concentration

The greater the concentration of enzyme, the more frequent the collisions between enzyme and substrate, and therefore the faster the rate of the reaction. However, at very high enzyme concentrations, the concentration of substrate may become a limiting factor, so the rate does not continue to increase if the enzyme concentration is increased.

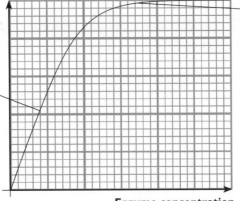

When there are more substrate molecules than enzyme molecules, running the experiment again with more enzyme produces a higher rate of reaction. The rate of reaction is in proportion to the concentration of enzyme used.

When there are more enzyme molecules than substrate molecules, running the experiment again with more enzyme does not result in a higher rate of reaction as there are no spare substrate molecules for the enzyme to act on.

How enzyme concentration affects the rate of an enzyme-catalysed reaction

Investigating the effect of enzyme concentration on rate of reaction

You could use the following method to investigate the effect of enzyme concentration on the rate at which the enzyme catalase converts its substrate, hydrogen peroxide, to water and oxygen.

Prepare a catalase solution as described on pages 36–37.

Prepare different dilutions of this solution:

Volume of initial solution/cm³	Volume of distilled water added/cm³	Relative concentration of catalase (as a percentage of the concentration of the initial solution)
10	0	100
9	1	90
8	2	80

and so on. The final 'solution' prepared should be 10 cm³ of distilled water.

Place each solution into a tube fitted with a gas syringe (see page 37). Use relatively small tubes, so that there is not too much gas in the tube above the liquid, but leave space to add an equal volume of hydrogen peroxide solution at the next step. Ensure that each tube is labelled with a waterproof marker. If time and materials allow, prepare three sets of these solutions.

Place each tube in a water bath at 30 °C.

Take another set of tubes and add 10 cm³ of hydrogen peroxide solution to each one. The concentration of hydrogen peroxide must be the same in each tube. Stand these tubes in the same water bath.

Leave all the tubes for at least 5 minutes to allow them to come to the correct temperature. When ready, add the contents of one of the hydrogen peroxide tubes to the first enzyme tube. Mix thoroughly. Measure the volume of gas collected in the gas syringe after two minutes. If you are using three sets, then repeat using the other two tubes containing the same concentration of enzyme.

Do the same for each of the tubes of enzyme. Record the mean volume of gas produced in two minutes for each enzyme concentration and plot a line graph to display your results.

Note: if you find that you get measurable volumes of gas sooner than two minutes after mixing the enzyme and substrate, then take your readings earlier. The closer to the start of the reaction you make the measurements, the better.

Substrate concentration

The greater the concentration of substrate, the more frequent the collisions between enzyme and substrate, and therefore the faster the rate of the reaction. However, at high substrate concentrations, the concentration of enzyme may become a limiting factor, so the rate does not continue to increase if the substrate concentration is increased.

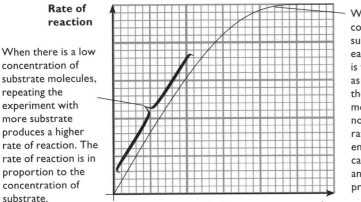

Rate of reaction

When there is a low concentration of substrate molecules, repeating the experiment with more substrate produces a higher rate of reaction. The rate of reaction is in proportion to the concentration of substrate.

When there is a high concentration of substrate molecules, each enzyme molecule is working as fast as it can. Repeating the experiment with more substrate does not result in a higher rate of reaction, as enzyme molecules cannot bind substrate and change it to product any faster.

Substrate concentration

How substrate concentration affects the rate of an enzyme-catalysed reaction

Investigating the effect of substrate concentration on the rate of an enzyme-catalysed reaction

You can do this in the same way as described for investigating the effect of enzyme concentration, but this time keep the concentration of catalase the same and vary the concentration of hydrogen peroxide.

Inhibitors

An inhibitor is a substance that slows down the rate at which an enzyme works.

Competitive inhibitors generally have a similar shape to the enzyme's normal substrate. They can fit into the enzyme's active site, preventing the substrate from binding. The greater the proportion of inhibitor to substrate in the mixture, the more likely it is that an inhibitor molecule, and not a substrate molecule, will bump into an active site. The degree to which a competitive inhibitor slows down a reaction is therefore affected by the relative concentrations of the inhibitor and the substrate.

Non-competitive inhibitors do not have the same shape as the substrate, and they do not bind to the active site. They bind to a different part of the enzyme. This changes the enzyme's shape, including the shape of the active site, so the substrate can no long bind with it. Even if you add more substrate, it still won't be able to bind, so the degree to which a non-competitive inhibitor slows down a reaction is **not** affected by the relative concentrations of the inhibitor and the substrate.

Enzyme acting on its normal substrate

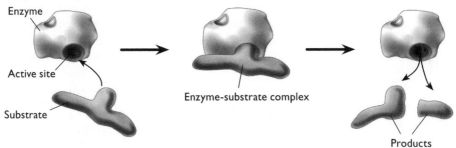

Enzyme in the presence of a competitive inhibitor

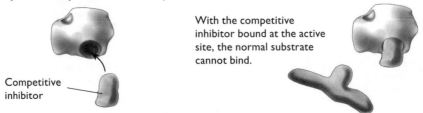

Enzyme in the presence of a non-competitive inhibitor

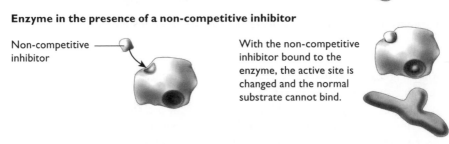

Competitive and non-competitive enzyme inhibitors

D Cell membranes and transport

Cell membranes

Every cell is surrounded by a cell membrane. There are also many membranes within cells. The membrane around the outside of a cell is called the cell surface membrane or plasma membrane.

Structure of a cell membrane

A cell membrane consists of a double layer of **phospholipid** molecules (page 28). This structure arises because in water a group of phospholipid molecules arranges itself into a **bilayer**, with the hydrophilic heads facing outwards into the water and the hydrophobic tails facing inwards, therefore avoiding contact with water.

This is the basic structure of a cell membrane. There are also **cholesterol** molecules in among the phospholipids. **Protein** molecules float in the phospholipid bilayer. Many of the phospholipids and proteins have short chains of carbohydrates attached to them, on the outer surface of the membrane. They are known as **glycolipids** and **glycoproteins**. There are also other types of glycolipid with no phosphate groups.

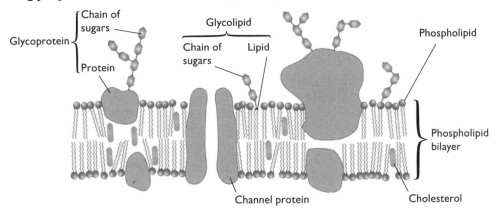

The fluid mosaic model of membrane structure

This is called the **fluid mosaic model** of membrane structure:
- 'fluid' because the molecules within the membrane can move around within their own layers
- 'mosaic' because the protein molecules are dotted around within the membrane
- 'model' because no-one has ever seen a membrane looking like the diagram — the molecules are too small to see even with the most powerful microscope. The structure has been worked out because it explains the behaviour of membranes that has been discovered through experiment.

The roles of the components of cell membranes

Component	Roles
Phospholipids	Form the fluid bilayer that is the fundamental structure of the membrane. Prevent hydrophilic substances — such as ions and some molecules — from passing through.
Cholesterol	Helps to keep the cell membrane fluid.
Proteins and glycoproteins	Provide channels that allow hydrophilic substances to pass through the membrane; these channels can be opened or closed to control the substances' movement. Actively transport substances through the membrane against their concentration gradient, using energy derived from ATP. Act as receptor molecules for substances such as hormones, which bind with them; this can then affect the activity of the cell. Cell recognition — cells from a particular individual or a particular tissue have their own set of proteins and glycoproteins on their outer surfaces.
Glycolipids	Cell recognition and adhesion to neighbouring cells to form tissues.

Passive transport through cell membranes

Molecules and ions are in constant motion. In gases and liquids they move freely. As a result of their random motion, each type of molecule or ion tends to spread out evenly within the space available. This is **diffusion**. Diffusion results in the net movement of ions and molecules from a high concentration to a low concentration.

Diffusion across a cell membrane

Some molecules and ions are able to pass through cell membranes. The membrane is permeable to these substances. However, some substances cannot pass through cell membranes, so the membranes are said to be **partially permeable**.

For example, oxygen is often at a higher concentration outside a cell than inside, because the oxygen inside the cell is being used up in respiration. The random motion of oxygen molecules inside and outside the cell means that some of them 'hit' the cell surface membrane. More of them hit the membrane on the outside than the inside, because there are more of them outside. Oxygen molecules are small and do not carry an electrical charge, so they are able to pass freely through the phospholipid bilayer. Oxygen therefore diffuses from outside the cell, through the membrane, to the inside of the cell, down its concentration gradient.

This is passive transport, because the cell does not do anything to cause the oxygen to move across the cell membrane.

Facilitated diffusion

Ions or electrically charged molecules are not able to diffuse through the phospholipid bilayer because they are repelled from the hydrophobic tails. Large molecules are also unable to move through the phospholipid bilayer freely. However, the cell membrane contains special protein molecules that provide hydrophilic passageways through which these ions and molecules can pass. They are called **channel proteins**. Different channel proteins allow the passage of different types of molecules and ions. Diffusion through these channel proteins is called **facilitated diffusion**. Like 'ordinary' diffusion, it is entirely passive.

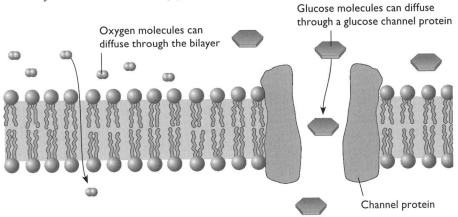

Diffusion across a cell membrane

Osmosis

Water molecules are small. They carry tiny electrical charges (dipoles) but their small size means that they are still able to move quite freely through the phospholipid bilayer of most cell membranes. Water molecules therefore tend to diffuse down their concentration gradient across cell membranes.

Cell membranes always have a watery solution on each side. These solutions may have different concentrations of solutes.

The greater the concentration of solute, the less water is present. The water molecules in a concentrated solution are also less free to move, because they are attracted to the solute molecules (see the diagram on page 48). A concentrated solution is therefore said to have a **low water potential**.

In a dilute solution, there are more water molecules and they can move more freely. This solution has a **high water potential**.

Imagine a cell membrane with a dilute solution on one side and a concentrated solution on the other side. The solute has molecules that are too large to get through the membrane — only the water molecules can get through.

Water molecules in the dilute solution are moving more freely and therefore hit the membrane more often than water molecules in the concentrated solution. More water molecules therefore diffuse across the membrane from the dilute to the concentrated solution than in the other direction. The net movement of water molecules is from a high water potential to a low water potential, down a water potential gradient.

This is **osmosis**. Osmosis is the diffusion of water molecules from a dilute solution to a concentrated solution through a partially permeable membrane, down a water potential gradient.

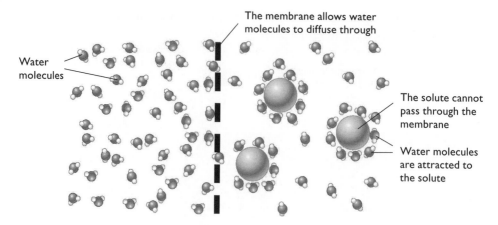

Osmosis across a cell membrane

Water potential is measured in pressure units, kilopascals (kPa). Pure water has a water potential of 0 kPa. Solutions have negative water potentials. For example, a dilute sucrose solution might have a water potential of –250 kPa. A concentrated sucrose solution might have a water potential of –4000 kPa. The more negative the number, the lower the water potential. Water moves by osmosis down a water potential gradient, from a high (less negative) water potential to a low (more negative) water potential.

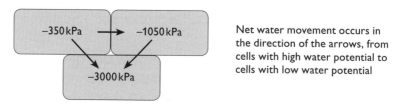

Net water movement occurs in the direction of the arrows, from cells with high water potential to cells with low water potential

Water movement between cells

Investigating the effect on plant cells of immersion in solutions of different water potentials

There are several different ways in which this investigation could be done. They include:

- cut cylinders or discs or strips of a solid and uniform plant tissue, such as a potato tuber, then measure either their lengths or masses before immersing them in the solutions. Leave them long enough for them to come to equilibrium, and then measure the length or mass of each piece again, and calculate the percentage change in the measurement. Percentage change can then be plotted against the concentration of the solution.

- cut small pieces of single-cell-thick plant tissue, for example onion epidermis. Mount them in a drop of sugar solution on microscope slides, and count the percentage of cells plasmolysed, or score each cell you see according to how plasmolysed it is.

For the plasmolysis investigation, you could use this method:

- Peel off one of the thick layers from an onion bulb. Cut 6 approximately 1 cm^2 pieces from it. Put each piece into a different liquid. One should be distilled water, then a range of sucrose solutions from about 0.1 mol dm^{-3} up to about 1.0 mol dm^{-3}.
- Take six clean microscope slides and label each with the concentration of solution you are going to place on it. Put a drop of the relevant solution on each one.
- Peel off the inner epidermis from one of the 1 cm^2 pieces of onion, and place it carefully onto the drop on a slide of the same concentration solution in which it has been immersed. Take care not to let it fold over. Press it gently into the liquid using a section lifter or other blunt tool. Carefully place a coverslip over it, taking care not to trap air bubbles.
- Repeat with each of the other drops of liquid. Leave all of the slides for at least 5 minutes to give any water movements by osmosis time to take place and equilibrium to be reached.
- Observe each slide under the microscope. Count the total number of cells in the field of view and record this. Then count the number that are plasmolysed and record this. Then move to another area of the slide and repeat until you have counted at least 50 cells. Repeat for each slide.
- Calculate the percentage of cells that have plasmolysed in each solution. Add this to your results chart.
- Plot percentage of cells that have plasmolysed (y-axis) against the concentration of the solution (x-axis). Join the points with either straight lines drawn between points, or a best-fit curve.
- The point at which the line crosses the 50% plasmolysis level tells you the concentration of the solution at which, on average, cells were just beginning to plasmolyse. At this value, the concentration of the solution inside the onion cells was, on average, the same as the concentration of the sucrose solution.

Active transport across cell membranes

low → high conc grad

Cells are able to make some substances move across their membranes *up* their concentration gradients. For example, if there are more potassium ions inside the cell than outside the cell, the potassium ions would diffuse out of the cell. However, the cell may require potassium ions. It may therefore use a process called **active**

transport to move potassium ions from outside the cell to inside the cell, against the direction in which they would naturally diffuse.

This is done using carrier (transporter) proteins in the cell membrane. These use energy from the breakdown of ATP to move the ions into the cell. The carrier proteins are ATPases.

$$\text{ATP} \xrightarrow{\text{ATPase}} \text{ADP + phosphate + energy}$$

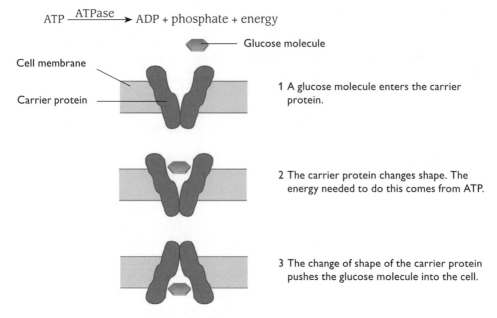

Active transport

Each carrier protein is specific to just one type of ion or molecule. Cells contain many different carrier proteins in their membranes.

Endocytosis and exocytosis

Cells can move substances into and out of the cell without the substances having to pass through the cell membrane.

In **endocytosis** the cell puts out extensions around the object to be engulfed. The membrane fuses together around the object, forming a vesicle.

In **exocytosis** the object is surrounded by a membrane inside the cell to form a vesicle. The vesicle is then moved to the cell membrane. The membrane of the vesicle fuses with the cell membrane, expelling its contents outside the cell.

Exocytosis

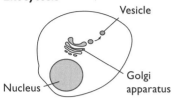

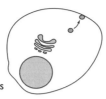

A vesicle is produced containing material to be removed from the cell.

The vesicle moves to the cell surface membrane.

The membrane of the vesicle and the cell surface membrane join and fuse.

The contents of the vesicle are released.

Endocytosis

The cell surface membrane grows out.

The object or solution is surrounded.

The membrane breaks and rejoins, enclosing the object and forming a vesicle.

The vesicle moves inwards and its contents are absorbed into the cytoplasm.

Endocytosis and exocytosis

Summary of methods of movement across membranes

Feature	Passive movement			Active movement	
	Diffusion	Osmosis	Facilitated diffusion	Active transport	Endocytosis and exocytosis
Requires energy input from the cell	No	No	No	Yes	Yes
Involves the movement of individual ions or molecules	Yes	Yes	Yes	Yes	No
Movement is through protein channels or protein carriers in the membrane	No	No	Yes	Yes	No
Examples of substances that move	Oxygen, carbon dioxide	Water	Ions (e.g. K^+, Cl^-) and molecules, e.g. glucose	Mostly ions (e.g. K^+, Cl^-)	Droplets of liquid; bacteria; export proteins

E Cell and nuclear division
Mitosis

A multicellular organism begins as a single cell. That cell divides repeatedly to produce all the cells in the adult organism.

The type of cell division involved in growth is called **mitosis**. Mitosis is also used to produce new cells to replace ones that have been damaged — that is, to repair tissues. Mitosis is also involved in asexual reproduction, in which a single parent gives rise to genetically identical offspring.

Strictly speaking, mitosis is division of the nucleus of the cell. After this, the cell itself usually divides as well. This is called **cytokinesis**.

The cell cycle

During growth of an organism many of its cells go through a continuous cycle of growth and mitotic division called the **cell cycle**.

The two G phases and S phase make up interphase.

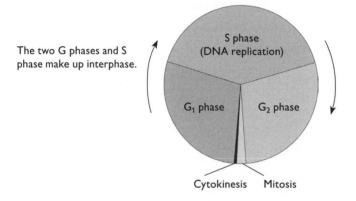

The cell cycle

For most of the cell cycle, the cell continues with its normal activities. It also grows, as the result of the production of new molecules of proteins and other substances, which increase the quantity of cytoplasm in the cell.

DNA replication takes place during interphase, so that there are two identical copies of each DNA molecule in the nucleus. (DNA replication is described on page 57). Each original chromosome is made up of one DNA molecule, so after replication is complete each chromosome is made of two identical DNA molecules. They are called **chromatids** and they remain joined together at a point called the **centromere**.

AS content guidance

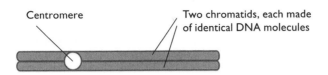

A chromosome before cell division

During mitosis, the two chromatids split apart and are moved to opposite ends of the cell. A new nuclear envelope then forms around each group. These two nuclei each contain a complete set of DNA molecules identical to those in the original (parent) cell. Mitosis produces two genetically identical nuclei from one parent nucleus (see diagram on page 54).

After mitosis is complete, the cell usually divides into two, with one of the new nuclei in each of the two new cells. The two daughter cells are genetically identical to each other and their parent cell.

Control of cell division

Each cell contains genes that help to control when it divides. It is important that cells divide by mitosis only when they are required to do so. This usually involves signals from neighbouring cells, to which the cell responds by either dividing or not dividing. If this control goes wrong, then cells may not divide when they should (so growth does not take place, or wounds do not heal) or they may divide when they should not (so that a tumour may form).

Cancer is a disease that can result when genes that normally control cell division mutate. The cell may divide over and over again, forming an irregular mass of cells. If the tumour is malignant, then some of these cells may break off and start to form new tumours elsewhere in the body.

It is thought that several different control genes must mutate before a cell becomes cancerous. This can happen just by chance. The risk is increased by any factor that can cause mutation, including:
- ionising radiation (from radioactive sources emitting α, β or γ radiation, and from X rays)
- ultraviolet radiation (in sunlight)
- various chemicals, including several contained in tar from tobacco smoke
- some viruses, for example the human papilloma virus (HPV), which can cause cervical cancer.

↗ *carcinogens*

Haploid and diploid cells

Most of the cells in your body are **diploid** cells. This means that they contain two complete sets of chromosomes. Each cell has 2 sets of 23 chromosomes, making 46 chromosomes altogether.

Prophase

Centrioles

- The chromosomes condense
- The centrioles duplicate
- The centriole pairs move towards each pole
- The spindle begins to form

Metaphase

Spindle fibre

- The nuclear envelope disappears
- The centriole pairs are at the poles
- The spindle is completely formed
- The chromosomes continue to condense
- The spindle fibres attach to the centromeres of the chromosomes
- The spindle fibres pull on the centromeres, arranging them on the equator

Anaphase

- The links between sister chromatids break
- The centromeres of sister chromatids move apart, pulled by the spindle fibres

Telophase

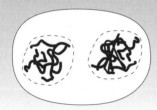

- Sister chromatids (now effectively separate chromosomes) reach opposite poles
- The chromosomes decondense
- Nuclear envelopes begin to form around the chromosomes at each pole
- The spindle disappears

Cytokinesis

- The cell divides into two cells, either by infolding of the plasma membrane in animal cells, or by the formation of a new cell wall and plasma membrane in plants

Mitosis and cytokinesis in an animal cell

In the original cell from which you began, one of these sets came from your father and one from your mother. As this cell divided by mitosis, each daughter cell obtained a complete copy of each set.

The sperm and egg that fused at fertilisation to produce that original cell each contained only *one* set of chromosomes. They were **haploid** cells. They each contained 23 chromosomes. When they fused together, this produced a diploid zygote with two sets of chromosomes.

Sperm and egg cells are produced from diploid cells by a special type of nuclear division called **meiosis**. In meiosis, the chromosomes are shared out so that each daughter cell gets only half of the original number of chromosomes. Meiosis produces haploid cells from diploid cells. In humans, meiosis only happens in the testes and ovaries. Meiosis is sometimes known as **reduction division**, because it reduces the number of chromosomes in a cell by half.

You do not need to know how meiosis takes place for the AS examination, only for A2. It is described on pages 172–175.

F Genetic control
DNA and RNA

DNA and RNA are **polynucleotides**. Polynucleotides are substances whose molecules are made of long chains of nucleotides linked together. A nucleotide is made up of:
- a 5-carbon sugar (**deoxyribose** in DNA; **ribose** in RNA)
- a phosphate group
- a nitrogen-containing base (**adenine**, **guanine**, **cytosine** or **thymine** in DNA; adenine, guanine, cytosine or **uracil** in RNA)

The bases are usually referred to by their first letters, A, G, C, T and U.

A and G are **purine** bases, made up of two carbon-nitrogen rings. C, T and U are **pyrimidine** bases, made up of one carbon-nitrogen ring.

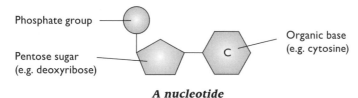

A nucleotide

Nucleotides can link together by the formation of covalent bonds between the phosphate group of one and the sugar of another. This takes place through a condensation reaction.

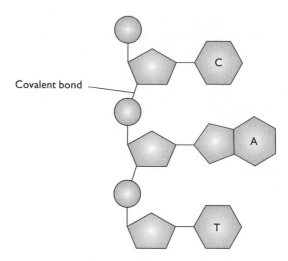

Part of a polynucleotide

An RNA molecule is usually made up of a single strand, although this may be folded up on itself. A DNA molecule is made up of two strands, held together by **hydrogen bonds** between the bases on the two strands. The strands run in opposite directions, i.e. they are anti-parallel.

Hydrogen bonding only occurs between A and T and between C and G. This is called **complementary base pairing**.

ATCG

A –T –(2)

C – G –(3)

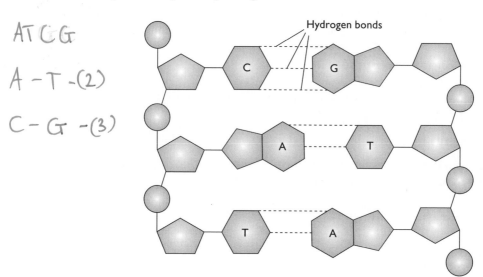

Part of a double-stranded DNA molecule

The two strands of nucleotides twist round each other to produce a double helix.

DNA replication

New DNA molecules need to be made before a cell can divide. The two daughter cells must each receive a complete set of DNA. The base sequences on the new DNA molecules must be identical with those on the original set. DNA replication takes place in the nucleus, during interphase.

- Hydrogen bonds between the bases along part of the two strands are broken. This 'unzips' part of the molecule, separating the two strands. — *enzyme helicase*
- Nucleotides that are present in solution in the nucleus are moving randomly around. By chance, a free nucleotide will bump into a newly exposed one with which it can form hydrogen bonds. Free nucleotides therefore pair up with the nucleotides on each of the DNA strands, always A with T and C with G.
- DNA polymerase links together the phosphate and deoxyribose groups of adjacent nucleotides.

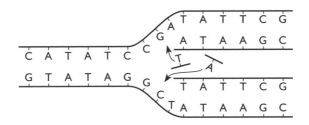

DNA replication

This is called **semi-conservative replication**, because each new DNA molecule has one old strand and one new one.

The genetic code

Genetic code determines sequence of amino acids

The sequence of bases in a DNA molecule is a code that determines the sequence in which amino acids are linked together when making a protein molecule. A sequence of DNA nucleotides that codes for one polypeptide, or for one protein, is known as a **gene**. */section of DNA*

As we have seen, the sequence of amino acids in a protein — its primary structure — determines its three-dimensional shape and therefore its properties and functions. For example, the primary structure of an enzyme determines the shape of its active site, and therefore the substrate with which it can bind.

A series of three bases in a DNA molecule, called a base **triplet**, codes for one amino acid. The DNA strand that is used in protein synthesis is called the **template**

strand. For example, this is the sequence of amino acids coded for by the template strand of a particular length of DNA:

bases in DNA	T A C	C T G	C A A	C T T
amino acid in polypeptide	methionine	aspartate	valine	glutamate

There are twenty amino acids. Because there are four bases, there are $4^3 = 64$ different possible combinations of bases in a triplet. Some amino acids therefore are coded for by more than one triplet. For example, the triplets AAA and AAG both code for the amino acid phenylalanine. The code is therefore said to be **degenerate**.

Protein synthesis

Proteins are made on the ribosomes in the cytoplasm, by linking together amino acids through peptide bonds. The sequence in which the amino acids are linked is determined by the sequence of bases on a length of DNA in the nucleus.

Transcription

The first step in protein synthesis is for the sequence of bases on the template strand of the DNA to be used to construct a strand of messenger RNA (**mRNA**) with a complementary sequence of bases. This is called **transcription**.

In the nucleus, the double helix of the DNA is unzipped, exposing the bases on each strand. There are four types of free RNA nucleotides in the nucleus, with the bases A, C, G and U. The RNA nucleotides form hydrogen bonds with the exposed bases on the template strand of the DNA. They pair up like this:

Base on DNA strand	Base on RNA strand
A	U
C	G
G	C
T	A

As the RNA nucleotides slot into place next to their complementary bases on the DNA, the enzyme **RNA polymerase** links them together (through their sugar and phosphate groups) to form a long chain of RNA nucleotides. This is an mRNA molecule.

The mRNA molecule contains a complementary copy of the base sequence on the template strand of part of a DNA molecule. Each triplet on the DNA is represented by a complementary group of three bases on the mRNA, called a **codon**.

1 Part of a molecule of DNA

2 The hydrogen bonds between bases are broken, exposing the bases

3 Free RNA nucleotides in the nucleus form new hydrogen bonds with the exposed bases on the template strand

4 The RNA nucleotides are linked together to form an mRNA molecule

mRNA molecule

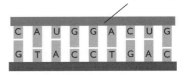

Transcription of part of a DNA molecule

Translation

The mRNA molecule breaks away from the DNA, and moves out of the nucleus into the cytoplasm. It becomes attached to a **ribosome**. Two codons fit into a groove in the ribosome. The first codon is generally AUG, which is known as a **start codon**. It codes for the amino acid methionine.

In the cytoplasm, 20 different types of amino acids are present. There are also many different types of transfer RNA (**tRNA**) molecules. Each tRNA molecule is made up of a single strand of RNA nucleotides, twisted round on itself to form a clover-leaf shape. There is a group of three exposed bases, called an **anticodon**. There is also a position at which a particular amino acid can be loaded by a specific enzyme.

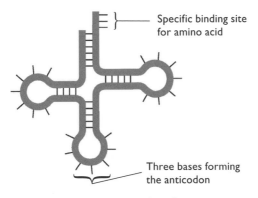

Specific binding site for amino acid

Three bases forming the anticodon

A tRNA molecule

The amino acid that can be loaded onto the tRNA is determined by the base sequence of its anticodon. For example, a tRNA whose anticodon is UAC will be loaded with the amino acid methionine.

A tRNA molecule with the complementary anticodon to the first codon on the mRNA, and carrying its appropriate amino acid, slots into place next to it in the ribosome, and hydrogen bonds form between the bases. Then a second tRNA does the same with the next mRNA codon.

H-bonds form between bases

The amino acids carried by the two adjacent tRNAs are then linked by a peptide bond.

peptide bonds form between amino acids

The mRNA is then moved along one place in the ribosome, and a third tRNA slots into place against the next mRNA codon. A third amino acid is added to the chain.

This continues until a **stop codon** is reached on the mRNA. This is a codon that does not code for an amino acid, such as UGA. The polypeptide (long chain of amino acids) that has been formed breaks away.

This process of building a chain of amino acids following the code on an mRNA molecule is called **translation**.

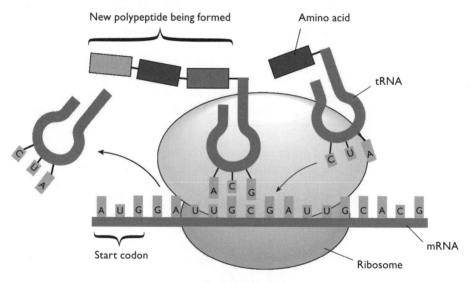

Translation

Mutation

A mutation is a random, unpredictable change in the DNA in a cell. It may be:
- a change in the sequence of bases in one part of a DNA molecule, or
- an addition of extra DNA to a chromosome or a loss of DNA from it, or
- a change in the total number of chromosomes in a cell.

Mutations are most likely to occur during DNA replication, for example when a 'wrong' base may slot into position in the new strand being built. Almost all of these mistakes are immediately repaired by enzymes, but some may persist. *change base sequence*

A change in the sequence of bases in DNA may result in a change in the sequence of amino acids in a protein. (Note that this does not always happen, because there is more than one triplet that codes for each amino acid, so a change in a triplet may not change the amino acid that is coded for.) This in turn may result in a change in the 3-D structure of the protein and therefore the way that it behaves.

↓ change in amino acid sequence ↓ change in 3D structure

Sickle cell anaemia

An example of a mutation is a change in the gene that codes for one of the polypeptides in a haemoglobin molecule. In the genetic disease **sickle cell anaemia**, the gene that codes for the β polypeptide has the base T where it should have the base A. This means that one triplet is different, so a different amino acid is used when the polypeptide chain is constructed on a ribosome.

These two different forms of the gene are called **alleles**.

Normal allele
DNA base sequence:
ATG GTG CAC CTG ACT CCT **GAG** GAG AAG TCT GCC GTT ACT
Amino acid sequence:
Val-His-Leu-Thr-Pro-**Glu**-Glu-Lys-Ser-

Abnormal allele
DNA base sequence:
ATG GTG CAC CTG ACT CCT **GTG** GAG AAG TCT GCC GTT ACT
Amino acid sequence:
Val-His-Leu-Thr-Pro-**Val**-Glu-Lys-Ser-

The abnormal β polypeptide has the amino acid valine where it should have the amino acid glutamic acid. These amino acids are on the outside of the haemoglobin molecule when it takes up its tertiary and quaternary shapes.

Glutamic acid is a hydrophilic amino acid. It interacts with water molecules, helping to make the haemoglobin molecule soluble.

Valine is a hydrophobic amino acid. It does not interact with water molecules, making the haemoglobin molecule less soluble.

When the abnormal haemoglobin is in an area of low oxygen concentration, the haemoglobin molecules stick to one another, forming a big chain of molecules that is not soluble and therefore forms long fibres. This pulls the red blood cells (inside which haemoglobin is found) out of shape, making them sickle-shaped instead of round. They are no longer able to move easily through the blood system and may get stuck in capillaries. This is very painful and can be fatal.

The normal form of haemoglobin is called HbA. The abnormal haemoglobin is called sickle cell haemoglobin (HbS), and the disease it causes is sickle cell anaemia.

G Transport
The need for transport systems

All cells need to take in substances from their environment, and get rid of unwanted substances. For example, a cell that is respiring aerobically has to take in oxygen and get rid of carbon dioxide.

In a single-celled organism, this can happen quickly enough by diffusion alone. This is because:
- no point in the cell is very far from the surface, so it does not take long for gases to diffuse from the cell surface membrane to the centre of the cell, or vice versa;
- the surface area to volume ratio of the cell is relatively large — that is, it has a large amount of surface area compared to its total volume.

In a large organism, diffusion is no longer sufficient. This is because:
- the centre of the organism may be a long way from the surface, so it would take too long for substances to diffuse all that way;
- the surface area to volume ratio is much smaller — that is, it has a small amount of surface area compared to its total volume.

Large organisms solve these difficulties in two ways:
- they have transport systems that carry substances by mass flow from one part of the body to another, rather than relying solely on diffusion;
- they increase the surface area of parts of the body involved in exchange with the environment, for example by having thin, flat leaves or by having a highly folded gas exchange surface.

Transport in plants

Plants can be very large, but they have a branching shape which helps to keep the surface area to volume ratio fairly large. Their energy needs are generally small compared with those of animals, so respiration does not take place so quickly. They can therefore rely on diffusion to supply their cells with oxygen and to remove carbon dioxide. Their leaves are very thin and have a large surface area inside them in contact with the air spaces. This means that diffusion is sufficient to supply the mesophyll cells with carbon dioxide for photosynthesis, and to remove oxygen.

Plant transport systems, therefore, do not transport gases. Plants have two transport systems:

xylem, which transports water and inorganic ions from the roots to all other parts of the plant;

phloem, which transports substances made in the plant, such as sucrose and amino acids, to all parts of the plant.

Transport in xylem

The diagram on page 64 shows the pathway taken by water through a plant.

The driving force that causes this movement is the loss of water vapour from the leaves. This is called **transpiration**.

Transpiration

Transpiration is the loss of water vapour from a plant. Most transpiration happens in the leaves. A leaf contains many cells in contact with air spaces in the mesophyll layers. Liquid water in the cell walls changes to water vapour, which diffuses into the air spaces. The water vapour then diffuses out of the leaf through the stomata, down a water potential gradient, into the air surrounding the leaf.

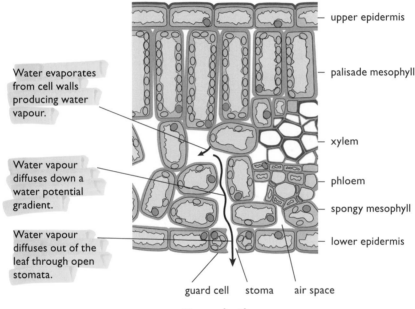

Water evaporates from cell walls producing water vapour.

Water vapour diffuses down a water potential gradient.

Water vapour diffuses out of the leaf through open stomata.

upper epidermis

palisade mesophyll

xylem

phloem

spongy mesophyll

lower epidermis

guard cell   stoma   air space

Transpiration

Each stoma is surrounded by a pair of guard cells. These can change shape to open or close the stoma. In order to photosynthesise, the stomata must be open so that carbon dioxide can diffuse into the leaf. Plants cannot therefore avoid losing water vapour by transpiration.

Transpiration is affected by several factors:
- High temperature increases the rate of transpiration. This is because at higher temperatures water molecules have more kinetic energy. Evaporation from the cell walls inside the leaf therefore happens more rapidly, and diffusion also happens more rapidly.
- High humidity decreases the rate of transpiration. This is because the water potential gradient between the air spaces inside the leaf and the air outside is less steep, so diffusion of water vapour out of the leaf happens more slowly.

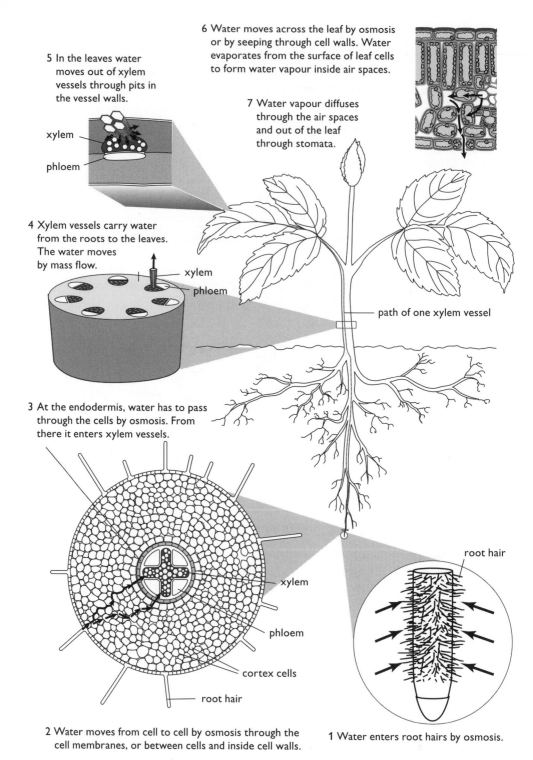

6 Water moves across the leaf by osmosis or by seeping through cell walls. Water evaporates from the surface of leaf cells to form water vapour inside air spaces.

5 In the leaves water moves out of xylem vessels through pits in the vessel walls.

xylem

phloem

7 Water vapour diffuses through the air spaces and out of the leaf through stomata.

4 Xylem vessels carry water from the roots to the leaves. The water moves by mass flow.

xylem

phloem

path of one xylem vessel

3 At the endodermis, water has to pass through the cells by osmosis. From there it enters xylem vessels.

root hair

xylem

phloem

cortex cells

root hair

2 Water moves from cell to cell by osmosis through the cell membranes, or between cells and inside cell walls.

1 Water enters root hairs by osmosis.

The pathway of water through a plant

- High wind speed increases the rate of transpiration. This is because the moving air carries away water vapour from the surface of the leaf, helping to maintain a water potential gradient between the air spaces inside the leaf and the air outside.
- High light intensity may increase the rate of transpiration. This is because the plant may be photosynthesising rapidly, requiring a rapid supply of carbon dioxide. This means that more stomata are likely to be open, through which water vapour can diffuse out of the leaf.

Investigating the factors that affect transpiration rate

It is difficult to measure the rate at which water vapour is lost from leaves. It is much easier to measure the rate at which a plant, or part of a plant, takes up water. Most of the water taken up is lost through transpiration, so we can generally assume that an increase in the rate of take-up of water indicates an increase in the rate of transpiration.

The apparatus used to measure the rate of take-up of water of a plant shoot is called a **potometer**. This can simply be a long glass tube. More complex potometers may have reservoirs which make it easier to refill the tube with water, or a scale marked on them.

- Fix a short length of rubber tubing over one end of the long glass tube. Completely submerge the tube in water. Move it around to get rid of all air inside it and fill it with water. Make absolutely sure there are no air bubbles.
- Take a leafy shoot from a plant and submerge it in the water alongside the glass tube. Using a sharp blade, make a slanting cut across the stem.
- Push the cut end of the stem into the rubber tubing. Make sure the fit is tight and that there are no air bubbles. If necessary, use a small piece of wire to fasten the tube tightly around the stem.
- Take the whole apparatus out of the water and support it upright. Wait at least ten minutes for it to dry out. If the glass tube is not marked with a scale, place a ruler or graph paper behind it.
- Start a stop clock and read the position of the air/water meniscus (which will be near the base of the tube). Record its position every 2 minutes (or whatever time interval seems sensible). Stop when you have 10 readings, or when the meniscus is one third of the way up the tube.
- Change the environmental conditions and continue to take readings. For example, you could use a fan to increase 'wind speed', or move the apparatus into an area where the temperature is higher or lower.
- Plot distance moved by meniscus against time for each set of readings, on the same axes. Draw best fit lines. Calculate the mean distance moved per minute, or calculate the slope of each line. This can be considered to be the rate of transpiration.

Xylem tissue

Xylem tissue contains dead, empty cells with no end walls. These are called **xylem vessel elements**. They are arranged in long lines to form **xylem vessels**. These are long, hollow tubes through which water moves by mass flow from the roots to all other parts of the plant.

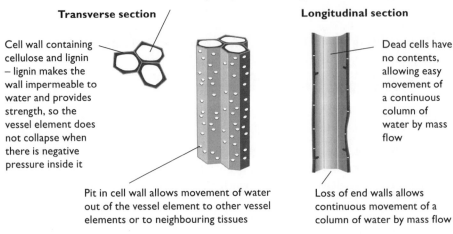

Narrow lumen increases area of water in contact with wall; water molecules adhere to the walls and this helps to prevent breakage of the water column

Transverse section

Longitudinal section

Cell wall containing cellulose and lignin – lignin makes the wall impermeable to water and provides strength, so the vessel element does not collapse when there is negative pressure inside it

Dead cells have no contents, allowing easy movement of a continuous column of water by mass flow

Pit in cell wall allows movement of water out of the vessel element to other vessel elements or to neighbouring tissues

Loss of end walls allows continuous movement of a column of water by mass flow

The structure of xylem vessels

How water moves from soil to air

Water moves from the soil to the air through a plant down a water potential gradient. The water potential in the soil is generally higher than in the air. The water potential in the leaves is kept lower than the water potential in the soil because of the loss of water vapour by transpiration. Transpiration maintains the water potential gradient.

- Water enters root hair cells by osmosis, moving down a water potential gradient from the water in the spaces between soil particles, through the cell surface membrane and into the cytoplasm and vacuole of the root hair cell.
- The water then moves from the root hair cell to a neighbouring cell by osmosis, down a water potential gradient. This is called the **symplast** pathway.
- Water also seeps into the cell wall of the root hair cell. This does not involve osmosis, as no partially permeable membrane is crossed. The water then seeps into and along the cell walls of neighbouring cells. This is called the **apoplast** pathway. In most plant roots, the apoplast pathway carries more water than the symplast pathway.
- When the water nears the centre of the root, it encounters a cylinder of cells called the **endodermis**. Each cell has a ring of impermeable **suberin** around it, forming the **Casparian strip**. This prevents water continuing to seep through cell walls. It therefore travels through these cells by the symplast pathway.
- The water moves into the xylem vessels from the endodermis.

- Water moves up the xylem vessels by mass flow — that is, in a similar way to water flowing in a river. The water molecules are held together by hydrogen bonds between them, keeping the water column unbroken. There is a relatively low hydrostatic pressure at the top of the column, produced by the loss of water by transpiration. This lowering of hydrostatic pressure causes a pressure gradient from the base to the top of the xylem vessel.
- In a leaf, water moves out of xylem vessels through pits, and then across the leaf by the apoplast and symplast pathways.
- Water evaporates from the wet cell walls into the leaf spaces, and then diffuses out through the stomata.

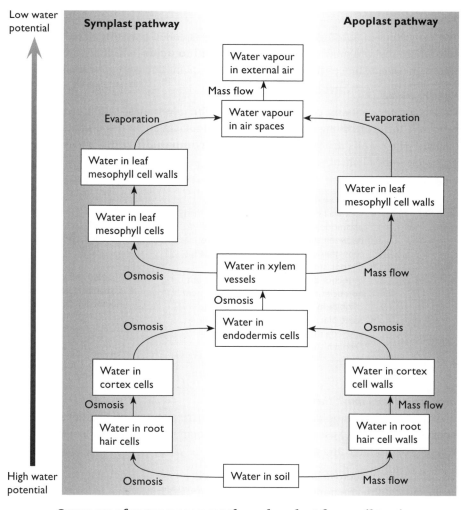

Summary of water movement through a plant from soil to air

Xerophytes

A xerophyte is a plant that is adapted to live in an environment where water is in short supply. The adaptations may include:

- leaves with a small surface area to volume ratio. This reduces the amount of surface area from which water vapour can diffuse.
- leaves with a thick, waxy cuticle. This reduces the quantity of water that can diffuse through the surface of the leaf into the air.
- methods of trapping moist air near the stomata, for example rolling leaf with stomata inside, having stomata in pits in the leaf surface, having hairs around the stomata. This produces a layer of high water potential around the stomata, reducing the water potential gradient and therefore reducing the rate of diffusion of water vapour from inside the leaf to outside.

Transport in phloem

The movement of substances in phloem tissue is called **translocation**. The main substances that are moved are **sucrose** and **amino acids**, which are in solution in water. These substances have been made by the plant and are called **assimilates**.

Phloem tissue

Phloem tissue contains cells called **sieve tube elements**. Unlike xylem vessel elements, these are living cells and contain cytoplasm and a few organelles but no nucleus. Their walls are made of cellulose. A **companion cell** is associated with each sieve tube element.

Longitudinal section

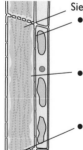

Sieve tube element with:
- Cell wall containing cellulose, with many plasmodesmata forming direct links between the cytoplasm of the sieve tube element and the companion cell.
- Cytoplasm containing some mitochondria and other organelles but no nucleus, leaving space for movement of phloem sap.
- Sieve plate — a perforated end wall allowing mass flow of phloem sap through the sieve pores.

Transverse section

Companion cell with:
- Cytoplasm containing numerous organelles, including a nucleus and many mitochondria.

Phloem tissue

How assimilates move through phloem tissue

A part of a plant where assimilates such as sucrose enter the phloem is called a **source**. A part where assimilates leave the phloem is called a **sink**. For example, a leaf may be a source and a root may be a sink.

Translocation of sucrose and other assimilates is an energy-requiring process.

- Respiration in companion cells at a source provides ATP that is used to fuel the active transport of sucrose into the companion cell. This increases the concentration of sucrose in the companion cell, so that it moves by diffusion down a concentration gradient into the phloem sieve element.

A. transport : sucrose → companion cell.

Diffusion : companion cell → phloem sieve element

SOURCE → SINK

- The increased concentration of sucrose in the companion cell and phloem sieve element produces a water potential gradient from the surrounding cells into the companion cell and phloem sieve element. Water moves down this gradient.
- At a sink, sucrose diffuses out of the phloem sieve element and down a concentration gradient into a cell that is using sucrose. This produces a water potential gradient, so water also diffuses out of the phloem sieve element.
- The addition of water at the source and the loss of water at the sink produces a higher hydrostatic pressure inside the phloem sieve element at the source than at the sink. Phloem sap therefore moves by **mass flow** down this pressure gradient, through the phloem sieve elements and through the sieve pores, from source to sink.

high hydrostatic pressure
at the source
low = sink

Transport in mammals

Mammals have a blood system made up of blood vessels and the heart. The heart produces high pressure, which causes blood to move through the vessels by mass flow. The blood system is called a **closed system** because the blood travels inside vessels. It is called a **double circulatory system** because the blood flows from the heart to the lungs, then back to the heart, then around the rest of the body and then back to the heart again. The vessels taking blood to the lungs and back make up the **pulmonary system**. The vessels taking blood to the rest of the body and back make up the **systemic system**.   *← systematic = blood to body.*

Blood vessels

Arteries carry blood away from the heart. The blood that flows through them is pulsing and at a high pressure. They therefore have thick, elastic walls which can expand and recoil as the blood pulses through. The artery wall also contains variable amounts of smooth muscle. Arteries branch into smaller vessels called **arterioles**. These also contain smooth muscle in their walls, which can contract and make the lumen (space inside) smaller. This helps to control the flow of blood to different parts of the body. (Note that the muscle in the walls of arteries does **not** help to push the blood through them.)

Capillaries are tiny vessels with just enough space for red blood cells to squeeze through. Their walls are only one cell thick, and there are often gaps in the walls through which plasma (the liquid component of blood) can leak out. Capillaries deliver nutrients, hormones and other requirements to body cells, and take away their waste products. Their small size and thin walls minimise diffusion distance, enabling exchange to take place rapidly between the blood and the body cells.   *rapid gas exchange*

Veins carry low-pressure blood back to the heart. Their walls do not need to be as tough or as elastic as those of arteries as the blood is not at high pressure and is not pulsing. The lumen is larger than in arteries, reducing friction which would otherwise slow down blood movement. They contain valves, to ensure that the blood does not flow back the wrong way. Blood is kept moving through many veins, for example those in the legs, by the squeezing effect produced by contraction of the body muscles close to them, which are used when walking.   *large lumen = less friction*

cytoskeleton/body muscles

69

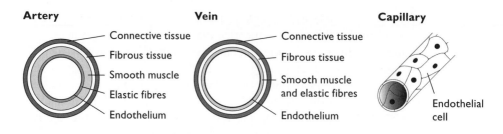

Artery

Connective tissue
Fibrous tissue
Smooth muscle
Elastic fibres
Endothelium

Vein

Connective tissue
Fibrous tissue
Smooth muscle
and elastic fibres
Endothelium

Capillary

Endothelial
cell

Structure of blood vessels

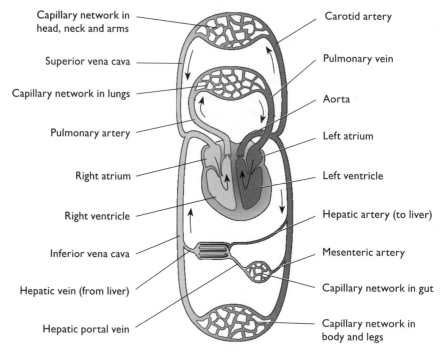

Capillary network in
head, neck and arms

Superior vena cava

Capillary network in lungs

Pulmonary artery

Right atrium

Right ventricle

Inferior vena cava

Hepatic vein (from liver)

Hepatic portal vein

Carotid artery

Pulmonary vein

Aorta

Left atrium

Left ventricle

Hepatic artery (to liver)

Mesenteric artery

Capillary network in gut

Capillary network in
body and legs

The main blood vessels in the human body

Pressure changes in the circulatory system

The pressure of the blood changes as it moves through the circulatory system.

- In the arteries, blood is at high pressure because it has just been pumped out of the heart. The pressure oscillates (goes up and down) in time with the heart beat. The stretching and recoil of the artery walls helps to smooth the oscillations, so the pressure becomes gradually steadier the further the blood moves along the arteries. The mean pressure also gradually decreases.
- The total cross-sectional area of the capillaries is greater than that of the arteries that supply them, so blood pressure is less inside the capillaries than inside arteries.
- In the veins, blood is at a very low pressure, as it is now a long way from the pumping effect of the heart.

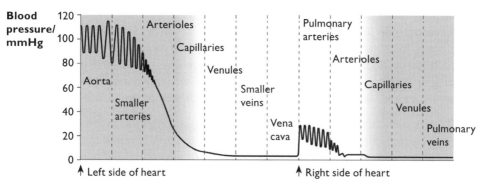

Pressure changes in the circulatory system

Blood

Blood contains the following cell types.

- **Red blood cells**. These transport oxygen from lungs to respiring tissues. They are very small and have the shape of a biconcave disc. This increases their surface area to volume ratio, allowing rapid diffusion of oxygen into and out of them. They contain haemoglobin, which combines with oxygen to form oxy-haemoglobin in areas of high concentration (such as the lungs) and releases oxygen in areas of low concentration (such as respiring tissues). They do not contain a nucleus or mitochondria.
- **Phagocytes**. These are white blood cells that engulf and digest unwanted cells, such as damaged body cells or pathogens. They are larger than red blood cells and often have a lobed nucleus.
- **Lymphocytes**. These are white blood cells that respond to particular pathogens by secreting antibodies or by directly destroying them. Each lymphocyte is able to recognise one particular pathogen and respond to it by secreting one particular type of antibody or by attacking it. You can find out more about this on pages 87–90.

Red blood cells — flattened discs, thinnest at the middle.

Platelets — small structures usually stained blue.

White blood cells contain a nucleus which is usually stained dark blue when a blood smear is prepared for viewing under a microscope. Several forms of white cells can be recognised, such as lymphocytes, monocytes and granulocytes.

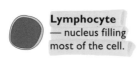

Lymphocyte — nucleus filling most of the cell.

Monocyte — the largest white cell, often with a kidney-shaped nucleus.

Granulocyte — has a lobed nucleus (often 3–5 lobes) and granules in the cytoplasm that may be stained pink or blue.

Blood cells

71

Haemoglobin and oxygen transport

Haemoglobin (Hb) is a protein with quaternary structure. A haemoglobin molecule is made up of four polypeptide chains, each of which has a **haem** group at its centre. Each haem group contains an Fe^{2+} ion which is able to combine reversibly with oxygen, forming **oxyhaemoglobin**. Each iron ion can combine with 2 oxygen atoms, so one Hb molecule can combine with 8 oxygen atoms.

Oxygen concentration can be measured as partial pressure, in kilopascals (kPa). Haemoglobin combines with more oxygen at high partial pressures than it does at low partial pressures. At high partial pressures of oxygen, all the haemoglobin will be combined with oxygen, and we say that it is 100% saturated with oxygen. A graph showing the relationship between the partial pressure of oxygen and the percentage saturation of haemoglobin with oxygen is known as a dissociation curve.

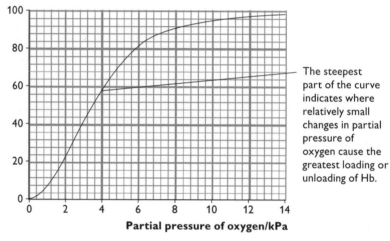

The steepest part of the curve indicates where relatively small changes in partial pressure of oxygen cause the greatest loading or unloading of Hb.

The oxygen dissociation curve for haemoglobin

In the lungs, the partial pressure of oxygen may be around 12 kPa. You can see from the graph that the Hb will be about 98% saturated.

In a respiring muscle, the partial pressure of oxygen may be around 2 kPa. The Hb will be about 23% saturated.

Therefore, when Hb from the lungs arrives at a respiring muscle it gives up more than 70% of the oxygen it is carrying.

The Bohr effect

The presence of carbon dioxide increases acidity, that is the concentration of H^+ ions. When this happens, the haemoglobin combines with H^+ ions and releases oxygen. Therefore, in areas of high carbon dioxide concentration, haemoglobin is less saturated with oxygen than it would be if there was no carbon dioxide present. This is called the **Bohr effect**. It is useful in enabling haemoglobin to unload more of its oxygen in tissues where respiration (which produces carbon dioxide) is taking place.

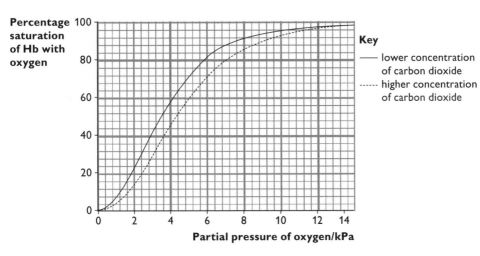

Key

—— lower concentration of carbon dioxide

······ higher concentration of carbon dioxide

The Bohr effect

Adaptation to high altitude

At high altitudes, the air is less dense and the partial pressure of oxygen is lower than at sea level. Haemoglobin is therefore less saturated with oxygen in the lungs and delivers less oxygen to body tissues.

After some time at high altitude, the number of red blood cells in the blood increases. This means that there are more haemoglobin molecules in a given volume of blood. Therefore, even though each Hb molecule carries less oxygen on average than at sea level, the fact that there are more of them helps to supply the same amount of oxygen to respiring tissues.

Athletes may make use of this by training at high altitude before an important competition. When they return to low altitude, their extra red blood cells can supply oxygen to their muscles at a greater rate than in an athlete who has not been to high altitude, giving them a competitive advantage.

Tissue fluid and lymph

Capillaries have tiny gaps between the cells in their walls. Near the arteriole end of a capillary, there is relatively high pressure inside the capillary, and plasma leaks out through these gaps to fill the spaces between the body cells. This leaked plasma is called **tissue fluid**.

Tissue fluid is therefore very similar to blood plasma. However, very large molecules such as albumin (a protein carried in solution in blood plasma) and other plasma proteins cannot get through the pores and so remain in the blood plasma.

The tissue fluid bathes the body cells. Substances such as oxygen, glucose or urea can move between the blood plasma and the cells by diffusing through the tissue fluid.

Some tissue fluid moves back into the capillaries, becoming part of the blood plasma once more. This happens especially at the venule end of the capillary, where blood pressure is lower, producing a pressure gradient down which the tissue fluid can flow. However, some of the tissue fluid collects into blind-ending vessels called lymphatic vessels. It is then called **lymph**.

Lymphatic vessels have valves that allow fluid to flow into them and along them but not back out again. They carry the lymph towards the subclavian veins (near the collarbone) where it is returned to the blood. The lymph passes through **lymphatic glands** where white blood cells accumulate. Lymph therefore tends to carry higher densities of white blood cells than are found in blood plasma or tissue fluid.

The heart

The heart of a mammal has four chambers. The two **atria** receive blood, and the two **ventricles** push blood out of the heart. The atria and ventricle on the left side of the heart contain oxygenated blood, while those on the right side contain deoxygenated blood. The walls of the heart are made of **cardiac muscle**.

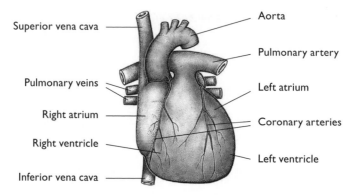

External view of a mammalian heart

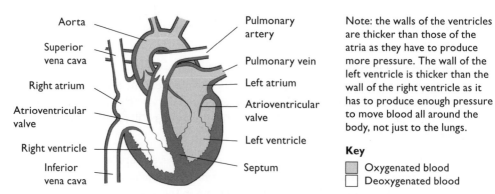

Vertical section through a mammalian heart

Note: the walls of the ventricles are thicker than those of the atria as they have to produce more pressure. The wall of the left ventricle is thicker than the wall of the right ventricle as it has to produce enough pressure to move blood all around the body, not just to the lungs.

Key

- ▨ Oxygenated blood
- ☐ Deoxygenated blood

The cardiac cycle

When muscle contracts, it gets shorter. Contraction of the cardiac muscle in the walls of the heart therefore causes the walls to squeeze inwards on the blood inside the heart. Both sides of the heart contract and relax together. The complete sequence of one heart beat is called the **cardiac cycle**.

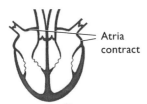 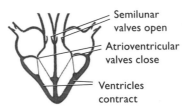 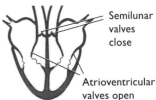

During **atrial systole**, the muscle in the walls of the atria contracts, pushing more blood into the ventricles through the open atrioventricular valves.

During **ventricular systole**, the muscle in the walls of the ventricles contracts. This causes the pressure of the blood inside the ventricles to become greater than in the atria, forcing the atrioventricular valves shut. The blood is forced out through the aorta and pulmonary artery.

During **diastole**, the heart muscles relax. The pressure inside the ventricles becomes less than that inside the aorta and pulmonary artery, so the blood inside these vessels pushes the semilunar valves shut. Blood flows into the atria from the veins, so the cycle is ready to begin again.

The cardiac cycle

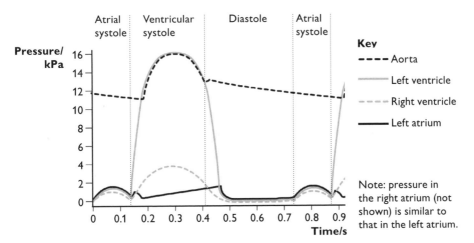

Pressure changes during the cardiac cycle

Initiation and control of the cardiac cycle

Cardiac muscle is **myogenic** — that is, it contracts and relaxes automatically, without the need of stimulation by nerves. The rhythmic, coordinated contraction of the cardiac muscle in different parts of the heart is coordinated through electrical impulses passing through the cardiac muscle tissue.

- In the wall of the right atrium, there is a patch of muscle tissue called the **sinoatrial node (SAN)**. This has an intrinsic rate of contraction a little higher than that of the rest of the heart muscle.
- As the cells in the SAN contract, they generate action potentials (electrical impulses — see page 166) which sweep along the muscle in the wall of the right and left atria. This causes the muscle to contract. This is atrial systole.
- When the action potentials reach the **atrioventricular node (AVN)** in the septum, they are delayed briefly. They then sweep down the septum between

the ventricles, along fibres of **Purkyne tissue**, and then up through the ventricle walls. This causes the ventricles to contract slightly after the atria. The left and right ventricles contract together, from the bottom up. This is ventricular systole.

- There is then a short delay before the next wave of action potentials is generated in the SAN. During this time, the heart muscles relax. This is diastole.

H Gas exchange

All organisms take in gases from their environment and release gases to the environment. Animals take in oxygen for aerobic respiration and release carbon dioxide. Plants also respire, but during daylight hours they photosynthesise at a greater rate than they respire, and so take in carbon dioxide and release oxygen.

The body surface across which these gases diffuse into and out of the body is called the gas exchange surface. In mammals, including humans, the gas exchange surface is the surface of the alveoli in the lungs.

The human gas exchange system

This is shown on page 77.

Cartilage in the walls of the trachea and bronchi provides support and prevents the tubes collapsing when the air pressure inside them is low.

Ciliated epithelium is found lining the trachea, bronchi and some bronchioles. It is a single layer of cells whose outer surfaces are covered with many thin extensions (cilia) which are able to move. They sweep mucus upwards towards the mouth, helping to prevent dust particles and bacteria reaching the lungs.

Goblet cells are also found in the ciliated epithelium. They secrete mucus, which traps dust particles and bacteria.

Smooth muscle cells are found in the walls of the trachea, bronchi and bronchioles. This type of muscle can contract slowly but for long periods without tiring. When it contracts, it reduces the diameter of the tubes. During exercise it relaxes, widening the tubes so more air can reach the lungs.

Elastic fibres are found in the walls of all tubes and between the alveoli. When breathing in, these fibres stretch to allow the alveoli and airways to expand. When breathing out, they recoil, helping to reduce the volume of alveoli and expel air out of the lungs.

Gas exchange at the alveolar surface

The air inside an alveolus contains a higher concentration of oxygen, and a lower concentration of carbon dioxide, than the blood in the capillaries. This blood has been brought to the lungs in the pulmonary artery, which carries deoxygenated blood from the heart. Oxygen therefore diffuses from the alveolus into the blood capillary, through the thin walls of the alveolus and the capillary. Carbon dioxide diffuses from the capillary into the blood.

The diffusion gradients for these gases are maintained by:
- breathing movements, which draw air from outside the body into the lungs, and then push it out again; this maintains a relatively high concentration of oxygen and low concentration of carbon dioxide in the alveoli;
- blood flow past the alveolus, which brings deoxygenated blood and carries away oxygenated blood.

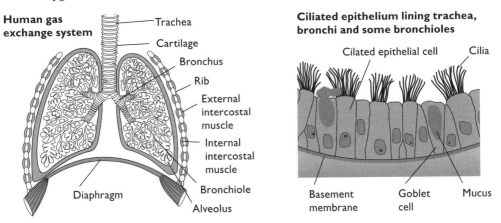

Cross sections of airways to show the distribution of tissues (not to scale)

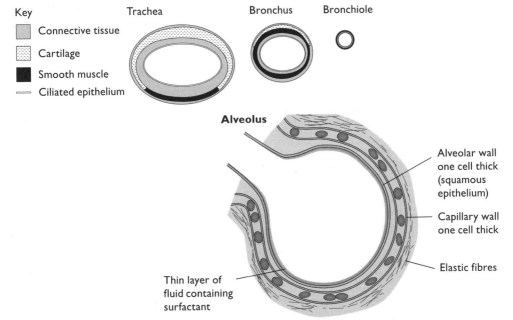

The structure of the human gas exchange system

Tidal volume and vital capacity

Air moves by mass flow into and out of the lungs during breathing. This is caused by the contraction and relaxation of external intercostal muscles and muscles in the diaphragm. When these contract, they increase the volume of the thoracic cavity and draw air down through the trachea and into the bronchi and bronchioles. When they relax, the thoracic volume decreases and air flows out, down a pressure gradient.

The volume of air that is moved into or out of the lungs during one breath is called the **tidal volume**. It is generally about 0.5 dm³. The maximum amount of air that can be moved in or out during the deepest possible breath is called the **vital capacity**. It is generally somewhere between 3 dm³ and 5 dm³.

Cigarette smoking and health

The smoke from cigarettes contains several substances that affect the gas exchange system and the cardiovascular system. These include:
- **tar**, a mixture of substances including various chemicals that act as **carcinogens**.
- **nicotine**, an addictive substance that affects the nervous system by binding to receptors on neurones (nerve cells) in the brain and other parts of the body. It increases the release of a neurotransmitter called dopamine in the brain, which gives feelings of pleasure. It increases the release of adrenaline into the blood, which in turn increases breathing rate and heart rate. There is also some evidence that nicotine increases the likelihood of blood clots forming.
- **carbon monoxide**, which combines irreversibly with haemoglobin, forming carboxyhaemoglobin. This reduces the amount of haemoglobin available to combine with oxygen, and so reduces the amount of oxygen that is transported to body tissues.

Effects of smoking on the gas exchange system

Chronic obstructive pulmonary disease (COPD)
This is a condition in which a person has chronic bronchitis and emphysema. It can be extremely disabling.

Chronic bronchitis
Various components of cigarette smoke, including tar, cause goblet cells to increase mucus production and cilia to beat less strongly. This causes mucus to build up, which may partially block alveoli. This makes gas exchange more difficult, as the diffusion distance between the air in the alveoli and the blood in the capillaries is greater. The mucus may become infected with bacteria, causing **bronchitis**. Smokers often have chronic (long-lasting) bronchitis.

The mucus stimulates persistent coughing, which can damage the tissues in the walls of the airways, making them stiffer and the airways narrower.

Emphysema

Smoking causes inflammation in the lungs. This involves the presence of increased numbers of white blood cells, some of which secrete chemicals that damage elastic fibres. This makes the alveoli less elastic. They may burst, resulting in larger air spaces. This reduces the surface area available for gas exchange. This is called **emphysema**. A person with emphysema has shortness of breath, meaning they struggle to breathe as deeply as they need to, especially when exercising.

Lung cancer

Various components of tar can cause changes in the DNA in body cells, including the genes that control cell division, which can cause cancer. These substances are therefore **carcinogens**. Cancers caused by cigarette smoke are most likely to form in the lungs but may form anywhere in the gas exchange system, and also in other parts of the body. Smoking increases the risk of developing all types of cancer. Symptoms of lung cancer include shortness of breath, a chronic cough — which may bring up blood — chest pain, fatigue and weight loss.

Effects of smoking on the cardiovascular system

The nicotine and carbon monoxide in tobacco smoke increase the risk of developing **atherosclerosis**. Atherosclerosis is a thickening and loss of elasticity in the walls of arteries. It is caused by build-up of plaques in the blood vessel wall. The plaques contain cholesterol and fibres. They produce a rough surface lining the artery, which stimulates the formation of blood clots.

A blood clot may break away from the artery wall and get stuck in a narrow vessel elsewhere in the blood system, for example in the lungs or in the brain. This prevents blood passing through so cells are not supplied with oxygen and die. If this happens in the brain it is called a **stroke**.

The loss of elasticity in an artery or arteriole also makes it more likely that the vessel will burst when high-pressure blood pulses through. This is another cause of stroke.

If atherosclerosis happens in the coronary arteries that supply the heart muscle with oxygenated blood, the person has **coronary heart disease** (CHD). Parts of the muscle may be unable to function properly as they do not have enough oxygen for aerobic respiration. The muscle may die. Eventually, this part of the heart may stop beating, causing a heart attack.

Evidence for effects of smoking on health

There are two ways in which the effects of smoking on health can be investigated.

Epidemiological evidence

This consists of data collected about people's smoking habits and their health. Large numbers of people should be involved in the study. The researchers then look for correlations between smoking and particular diseases. Although this approach does not provide any definite evidence about a causal link between smoking and the

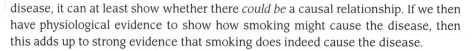

disease, it can at least show whether there *could be* a causal relationship. If we then have physiological evidence to show how smoking might cause the disease, then this adds up to strong evidence that smoking does indeed cause the disease.

Experimental evidence

This consists of carrying out controlled experiments. For example, the independent variable could be whether or not a subject smokes (or how much they smoke) and the dependent variable could be some aspect of physiology. All other variables should be kept constant. This is not possible with humans, as it would be unethical to make people smoke. In the 1960s, dogs and other animals were used in such experiments. The results showed conclusively that smoking tobacco greatly increases the risk of developing lung cancer. Experiments can also be carried out using cells grown in tissue culture. Exposure of these cells to chemicals found in tar shows that these chemicals can damage DNA.

Preventing and treating CHD

The risk of developing CHD is increased by:
- inheriting particular alleles of genes
- eating a diet rich in saturated fats and cholesterol
- not taking sufficient exercise
- being obese
- smoking

Severe CHD can be treated with a coronary bypass, in which a piece of blood vessel is taken from another part of the body and sewn into place to provide an alternative route for oxygenated blood to flow from the aorta to the heart muscle.

If the heart is damaged beyond repair, either by CHD or other conditions, then the only long-term option may be a heart transplant. The heart must come from a person who has just died (often in an accident) and has a tissue type that is similar to the recipient. Even so, the recipient will still have to take immunosuppressant drugs for the rest of their life, to prevent their immune system from attacking the donor tissues and rejecting the transplant.

Prevention of CHD and other forms of heart disease is clearly much better than having to carry out complex surgery. Lifestyle choices can be made that reduce the risks listed above (apart, of course, from the genes a person has). However, research shows that slightly obese people are more likely to recover well after heart surgery than thinner people.

I Infectious disease

Disease can be defined as a condition in which the body does not function normally, and which produces unpleasant symptoms such as pain, distress or feeling weak. The term disease is generally used for conditions that last for at least several days.

An **infectious disease** is one that can be passed between one person and another. Infectious diseases are caused by **pathogens**. These are usually microorganisms such as viruses, bacteria, fungi or protoctists.

A **non-infectious disease** cannot be passed between people and is not caused by pathogens. Examples include sickle cell anaemia and lung cancer.

Important infectious diseases

Cholera

Cause
Cholera is caused by a bacterium, *Vibrio cholerae*.

Transmission
V. cholerae can enter the body in contaminated food or water. The bacteria breed in the small intestine, where they secrete a toxin that reduces the ability of the epithelium of the intestine to absorb salts and water into the blood. These are lost in the faeces, causing diarrhoea. If not treated, the loss of fluid can be fatal. Cholera is most likely to occur where people use water or food that has been in contact with untreated sewage, as the bacteria are present in the faeces of an infected person.

Prevention and control
Transmission is most likely to occur in crowded and impoverished conditions, such as refugee camps. Cholera is best controlled by treating sewage effectively, providing a clean water supply and maintaining good hygiene in food preparation. There is no fully effective vaccine for cholera.

Biological control:
- Stocking ponds, irrigation
- spraying bacterium - kills
larvae.

Malaria

Cause
Malaria is caused by a protoctist, *Plasmodium*. There are several species, which cause different types of malaria. In a person, the *Plasmodium* infects red blood cells and breeds inside them. Toxins are released when the *Plasmodium* burst out of the cells, causing fever.

Transmission
Plasmodium is transmitted in the saliva of female *Anopheles* mosquitoes, which inject saliva to prevent blood clotting when they feed on blood from a person. When a mosquito bites an infected person, *Plasmodium* is taken up into the mosquito's body and eventually reaches its salivary glands. The mosquito is said to be a **vector** for malaria.

← anti - malarial drugs
eg quinine

Prevention and control
Reducing the population of mosquitoes, for example by removing sources of water in which they can breed, or by releasing large numbers of sterile males, can reduce the transmission of malaria.

1) reduce number of mosquitoes
2) avoid being bitten
3) use drugs to prevent parasite
infection.

prophylactic
(prevent infection
if person bitten)

Preventing mosquitoes from biting people, for example by sleeping under a mosquito net, or by wearing long-sleeved clothing and insect repellant, can reduce the chances of a mosquito picking up *Plasmodium* from an infected person, or passing it to an uninfected person.

Prophylactic drugs (that is, drugs that prevent pathogens infecting and breeding in a person) can be taken. However, in many parts of the world *Plasmodium* has evolved resistance to some of these drugs.

1) plasmodium became resistant to drugs used

Tuberculosis (TB)

2) mosquitoes became resistant to DDT – insecticides.

Cause

TB is caused by the bacterium *Mycobacterium tuberculosis* or (more rarely) *Mycobacterium bovis.*

bacteria can be inactive but later become active

Transmission

M. tuberculosis can enter the lungs in airborne droplets of liquid that are breathed in. This is more likely to happen in places where many people are living in crowded conditions.

Prevention and control ← *contact tracing*

TB causes HIV positives to die

TB is most prevalent amongst people living in poor accommodation, or whose immune systems are not functioning well, perhaps because of malnutrition or infection with HIV (see below). Increasing standards of living and treating HIV infection can therefore help to reduce the incidence of TB.

Vaccination with the BCG vaccine confers immunity to TB in many people. New vaccines are being developed that it is hoped will be more effective.

Treatment of HIV with drug therapy reduces the risk that an HIV-positive person will get TB.

streptomycin can inhibit

Treatment of TB with antibiotics can often completely cure the disease. However, this is not always the case because:

[MUST complete course]

- there are now many strains of the *M. tuberculosis* bacterium that have evolved resistance to most of the antibiotics that are used;
- the bacteria reproduce *inside* body cells, where it is difficult for drugs to reach them;
- the drugs need to be taken over a long time period, which often requires a health worker checking that a person takes their drugs every day.

HIV/AIDS

cannot survive outside human body

Cause

AIDS is caused by the human immunodeficiency virus, HIV. This is a retrovirus, which contains RNA. The virus enters T-lymphocytes (page 88), where its RNA is used to make viral DNA which is incorporated into the T-lymphocytes' chromosomes. Usually nothing more happens for several years, but eventually multiple copies of the virus are made inside the T-lymphocytes, which are destroyed as the viruses break out and infect more cells. Eventually there are so few functioning T-lymphocytes that the person is no longer able to resist infection by other pathogens and develops one or more opportunistic diseases such as TB.

Transmission

HIV can be passed from one person to another through:

- blood from one person entering that of another, for example by sharing hypodermic needles, or through blood transfusions
- exchange of fluids from the penis, vagina or anus
- across the placenta from mother to fetus, or in breast milk

Prevention and control

All blood to be used in transfusions should be screened to ensure it does not contain HIV.

All hypodermic needles should be sterile and used only once, and disposed of carefully.

A person should avoid sexual activity with anyone whose HIV status they do not know. If everyone had only one partner, HIV could not be transmitted. Condoms, if properly used, can prevent the virus passing from one person to another during intercourse. If a person is diagnosed with HIV, all their sexual contacts should be traced and informed that they may have the virus.

The chance of HIV passing from an HIV-positive mother to her fetus is greatly reduced if the mother is treated with appropriate drugs. These drugs can also greatly increase the length of time between a person becoming infected with HIV and developing symptoms of AIDS, and can significantly prolong life.

HIV infection rates are especially high in sub-Saharan Africa. Many of these people are not able to receive treatment with effective drugs, generally for economic reasons.

Smallpox

(virus changes its surface proteins ∴ hard to VACCINATE)

Cause

Smallpox is caused by the variola virus.

Transmission

Transmission occurs by the inhalation of droplets of moisture containing the virus.

Prevention and control

Smallpox was a serious disease that was fatal in 20–60% of adults who caught it, and in an even higher percentage of infected children. It was eradicated by 1979, through a vaccination campaign coordinated by the World Health Organization.

Measles

Cause

Measles is caused by a morbillivirus.

Transmission

Transmission occurs through the inhalation of droplets of moisture containing the virus. It is highly infectious, so that a high proportion of people who come into close contact with an infected person will also get the disease.

Prevention and control

Measles is a serious disease, which can cause death, especially in adults and in people who are not in good health, for example because they do not have access to a good diet. Vaccination is the best defence against measles. The vaccine is highly effective, especially if two doses are given. The people most likely to suffer from measles are therefore those who are malnourished and who live in areas where no vaccination programme is in place.

Global patterns of disease

Malaria is found in parts of the world where the *Anopheles* mosquito species that can act as vectors are found. This is mostly in tropical and subtropical regions where humidity is high.

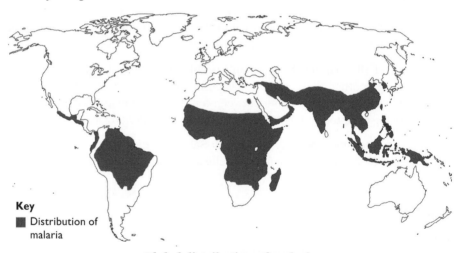

Key
■ Distribution of malaria

Global distribution of malaria

TB is found in all countries of the world, including developed countries such as the USA and the United Kingdom. However, it is most common in areas where living conditions are poor and crowded, or where large numbers of people have HIV/AIDS.

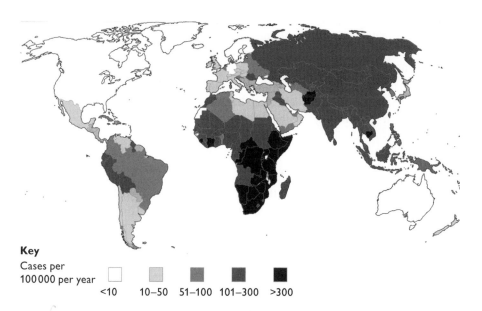

Key

Cases per 100 000 per year

    <10   10–50   51–100 101–300 >300

Global distribution of TB

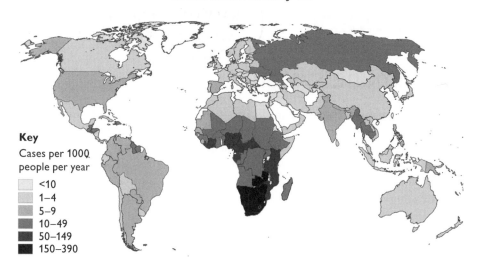

Key

Cases per 1000 people per year

- <10
- 1–4
- 5–9
- 10–49
- 50–149
- 150–390

Global distribution of HIV/AIDS

Antibiotics

An antibiotic is a substance that, when taken orally or by injection, kills bacteria but does not harm human cells. Antibiotics are not effective against viruses. Many antibiotics are originally derived from fungi, but they can also be obtained from other organisms (for example, amphibian skin or plants) or synthesised in the laboratory.

(treatment for
TB - isoniazid.)

Antibiotics act on structures or metabolic pathways that are found in bacteria but not in eukaryotic cells. The correct antibiotic must be chosen for a particular disease. For example:

- **penicillin** prevents the synthesis of the links between peptidoglycan molecules in bacterial cell walls; when the bacteria take up water by osmosis, the cell wall is not strong enough to prevent them bursting.
- **rifampicin (rifampin)** inhibits an enzyme required for RNA synthesis in bacteria.
- **tetracycline** binds to bacterial ribosomes and inhibits protein synthesis.

Exposure to antibiotics exerts strong selection pressure on bacterial populations. Any bacterium that is resistant to the antibiotic — for example, because it synthesises an enzyme that can break down the antibiotic — has a selective advantage and is more likely to survive and reproduce successfully. The offspring will inherit the alleles that confer resistance. A whole population of resistant bacteria can therefore be produced. For this reason, it is important that antibiotics are only used when necessary. A person prescribed antibiotics should complete the course, as this increases the chances of eradicating all the disease-causing bacteria in the body.

J Immunity

(bacteria have a single loop of double stranded DNA.)

The immune system

The human immune system is made up of the organs and tissues involved in destroying pathogens inside the body. There are two main groups of cells involved:
- phagocytes, which ingest and digest pathogens or infected cells;
- lymphocytes, which recognise specific pathogens through interaction with receptors in their cell surface membranes, and respond in one of several ways, for example by secreting antibodies.

Phagocytes

Phagocytes are produced in the bone marrow by the mitotic division of precursor cells. This produces cells that develop into **monocytes** or **neutrophils.**

Monocytes are inactive cells which circulate in the blood. They eventually leave the blood, often as the result of encountering chemical signals indicating that bacteria or viruses are present. As monocytes mature, they develop more RER, Golgi apparatus and lysosomes. When they leave the blood they become **macrophages**. They engulf bacteria by endocytosis (page 51) and digest them inside phagosomes. Monocytes and macrophages can live for several months.

Similar precursor cells in bone marrow produce **neutrophils**. These also travel in blood. They leave the blood in large numbers at sites of infection and engulf and digest bacteria in a similar way to macrophages. A neutrophil lives for only a few days.

Monocyte   Neutrophil

Mature phagocytes

Phagocytes are able to act against any invading organisms. Their response is non-specific.

The immune response

Lymphocytes, unlike phagocytes, act against specific pathogens. Each lymphocyte contains a set of genes that codes for the production of a particular type of receptor. We have many million different types, each producing just one type of receptor.

Both **B-lymphocytes** and **T-lymphocytes** are made in bone marrow. B-lymphocytes then spread through the body and settle in lymph nodes, although some continue to circulate in the blood. T-lymphocytes collect in the thymus gland, where they mature before spreading into the same areas as B-lymphocytes. The thymus gland disappears at around the time of puberty. Both types of lymphocyte have a large, rounded nucleus that takes up most of the cell. They can only be told apart by their different actions (see below).

During the maturation process, any lymphocytes that produce receptors that would bind with those on the body's own cells are destroyed. This means that the remaining lymphocytes will only act against **non-self** molecules that are not normally found in the body. Non-self molecules, such as those on the surfaces of invading bacteria, are called **antigens**.

Several different types of cell, including macrophages, place antigens of pathogens they have encountered in their cell surface membranes, where there is a good chance that a B-lymphocyte or T-lymphocyte may encounter them. These cells are called **antigen-presenting cells**.

Action of B-lymphocytes

A B-lymphocyte places some of its specific receptor molecules in its cell surface membrane. If it encounters an antigen that binds with this receptor, the B-lymphocyte is activated. It divides repeatedly by mitosis to produce a clone of genetically identical **plasma cells**. Some of these synthesise and secrete large quantities of proteins called **immunoglobulins** or **antibodies**. The antibodies have the same binding sites as the specific receptors in the B-lymphocyte's membrane, so they can bind with the antigens. This may directly destroy or neutralise the antigens, or it may make it easier for phagocytes to destroy them.

Some of the clone of B-lymphocyte cells become **memory cells**. These remain in the blood for many years. They are able to divide rapidly to produce plasma cells if the same antigen invades the body again. More antibody is therefore secreted more rapidly than when the first invasion happened, and it is likely that the pathogens will be destroyed before they have a chance to reproduce. The person has become **immune** to this pathogen.

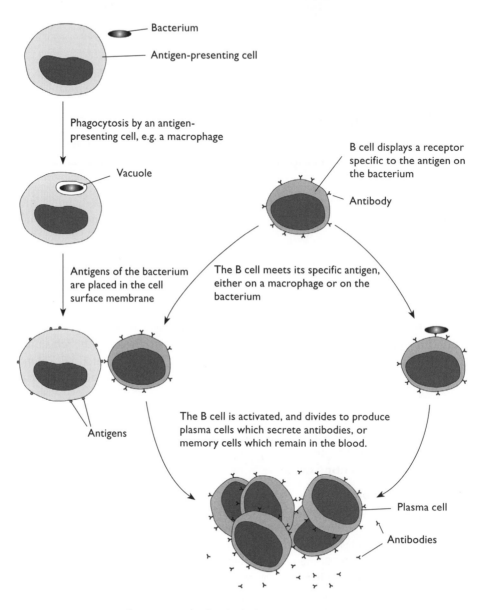

The response of B-lymphocytes to antigens

Action of T-lymphocytes

T-lymphocytes include **T helper cells** and **T killer cells**. Both of these types of cell place their specific receptors in their cell surface membranes. On encountering the relevant antigen, they are activated and divide by mitosis to form a clone.

Activated T helper cells secrete chemicals called **cytokines**. These stimulate B-lymphocytes to produce plasma cells, and stimulate monocytes and macrophages to attack and destroy pathogens.

Activated T killer cells attach to body cells that display the antigen matching their receptor. This happens when a body cell has been invaded by a virus. The T killer cell destroys the infected body cell.

Some of the clone of T cells become memory cells, which remain in the body and can react swiftly if the same pathogen invades again.

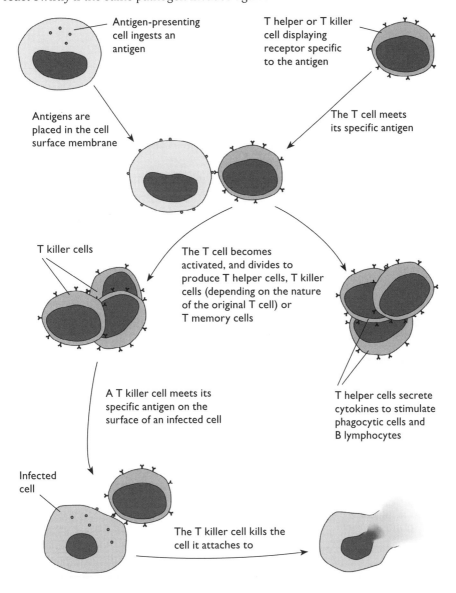

The response of T-lymphocytes to antigens

Antibodies

Antibodies are glycoproteins called **immunoglobulins** that are secreted by plasma cells in response to the presence of antigens.

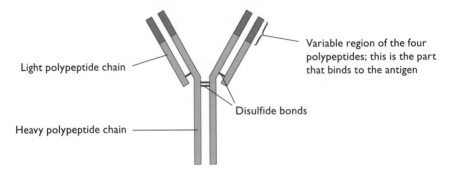

Light polypeptide chain

Variable region of the four polypeptides; this is the part that binds to the antigen

Disulfide bonds

Heavy polypeptide chain

An immunoglobulin molecule

The variable region of the immunoglobulin molecule is specific to the particular clone of B-lymphocytes that secreted it. It is able to bind with a particular type of antigen molecule. Immunoglobulins can:

- stick bacteria together, making it impossible for them to divide or making it easier for phagocytes to destroy them;
- neutralise toxins (poisonous chemicals) produced by pathogens;
- prevent bacteria from sticking to body tissues;
- bind to viruses and prevent them infecting cells.

Immunity

A person is immune to a disease if the pathogen that causes the disease is unable to reproduce in the body and cause illness. This happens when the body already contains, or is able rapidly to make, large quantities of antibodies against the antigens associated with the pathogen.

Active immunity occurs when the person's own lymphocytes make the antibody. This could be **natural**, as a result of the person having previously had the disease and forming B or T memory cells. It could also be **artificial**, as a result of **vaccination**. This involves introducing weakened pathogens into the body. The lymphocytes react to the antigens on the pathogens by producing antibodies and memory cells.

Passive immunity occurs when antibodies from elsewhere are introduced into the body. In a young baby this can be **natural**, as the baby acquires antibodies from its mother in breast milk. It can also be **artificial**, as the result of an injection of antibodies obtained from another animal.

Active immunity lasts much longer than passive immunity, because memory cells last a long time, whereas individual antibodies do not. Injections of antibodies, however, can be useful if a person requires instant immunity, for example if an aid worker is about to travel to an environment where risk of a disease such as hepatitis is high.

Global eradication of infectious disease

The World Health Organization has helped to organise world-wide campaigns to eliminate the serious infectious diseases smallpox and poliomyelitis.

Smallpox has been successfully eradicated by vaccinating large numbers of children with weakened viruses similar to those that cause smallpox. This succeeded because the vaccine is highly effective. The programme involved the vaccination of all relatives and contacts of anyone who had the disease, called ring vaccination. The virus did not mutate, so the same vaccine could be used everywhere.

Diseases that have *not* been successfully eradicated include:
- **Measles**. This is partly because several successive doses of vaccine are required to produce immunity, especially in children who have weak immune systems because of poor diet or living conditions. The virus is very infective, so a very high percentage of people must be vaccinated to ensure a population is free of it. Booster vaccinations are also needed. This is difficult to achieve in places where infrastructure is poor.
- **TB**. The BCG vaccination gives only partial immunity, although new vaccines are now being developed which it is hoped will be more effective. TB is difficult to treat because the bacteria live inside body cells. Many strains have developed resistance to antibiotics.
- **Malaria**. No effective vaccine has yet been developed against *Plasmodium*. This is a eukaryotic organism, not a bacterium or virus, and is not affected by antibodies produced by B-lymphocytes or by T-lymphocytes.
- **Cholera**. This disease is caused by the bacterium *Vibrio cholerae*. In the body, it lives and reproduces in the intestine, which is outside the body tissues and not easily reachable by lymphocytes or antibodies. Current cholera vaccines are ineffective, partly because injected vaccines do not readily reach the intestines. Oral vaccines are being developed, which are proving more effective.

Vaccines are, of course, completely ineffective against any diseases that are not caused by pathogens, such as sickle cell anaemia.

K Ecology

Ecology is the study of the ways in which organisms interact with their environment.

Levels of ecological organisation

A **habitat** is a type of environment in which an organism lives. For example, the habitat of a giraffe is grassland (savannah) with groups of trees such as *Acacia*. The habitat of a woodlouse (*Oniscus*) is a humid, dark place such as beneath the bark of a rotting log. The habitat of a mangrove tree is a muddy sea shore that is regularly flooded by the tide.

A **population** is a group of organisms of the same species that lives in the same place at the same time. If the species is a sexually-reproducing one, the organisms in the population are able to interbreed with one another. For example, all the giraffes in a particular area of savannah make up the giraffe population.

A **community** is all the organisms, of all the different species, that live in the same place at the same time. For example, all the giraffes and other animals, all the plants, all the fungi and all the bacteria make up a community in the savannah. Each type of habitat tends to have its own typical community.

An **ecosystem** is the interactions that take place between all the organisms in a community and their non-living environment. For example, an ecosystem in an area of African savannah would include the predator-prey relationships between giraffes and lions, the feeding relationships between grass and giraffes, the exchanges of oxygen and carbon dioxide between the air and the living organisms, the availability of mineral ions in the soil that can be taken up by plant roots, and so on. Strictly speaking, an ecosystem is not simply a *place* but a dynamic series of interactions between organisms and their environment.

A **niche** is the role of an organism in an ecosystem. Different species have different niches, although these may overlap. For example, both giraffes and zebras are herbivores that require open grassland and a water supply. However, giraffes are able to browse on vegetation from high tree branches, whereas zebras graze on grass and other low-growing plants.

Note that you are expected to have studied an ecosystem in an area familiar to you.

Energy flow through ecosystems

Living organisms require energy to maintain metabolic processes that keep their cells alive. Most of this energy is released from organic molecules such as glucose by respiration. The energy released is used to make ATP. The energy can then be released in small quantities, exactly when and where it is required, by hydrolysing the ATP to ADP and inorganic phosphate.

Each organism therefore needs a supply of energy-containing organic molecules in order to be able to make ATP. Organisms that can use energy from other sources, such as sunlight, to make these organic molecules are called **producers**. In most ecosystems, the producers are plants, which make carbohydrates by photosynthesis. They absorb energy from sunlight and incorporate it into carbohydrates, where it is stored as chemical potential energy.

Animals and fungi depend on taking in organic molecules that were originally synthesised by plants. They are **consumers**.

A **food chain** shows the pathway by which energy is passed from one organism to another. The energy is transferred in the form of chemical potential energy in food. The arrows in the food chain indicate the direction of energy transfer. A **food web** is a network of interconnecting food chains.

The position at which an organism feeds in a food chain is called a **trophic level.** Producers are at the first trophic level, primary consumers (herbivores) at the second trophic level, secondary consumers (carnivores that feed on herbivores) at the third trophic level, and so on.

Large quantities of energy are lost in the transfer between one trophic level and the next. For example, only about 10% of the energy in the grass in an area of savannah is passed on to herbivores. This is because:

- Not all the grass is eaten. Some is trampled or covered by animal droppings, or may grow too low to the ground for animals to be able to graze it. Pollen from grass flowers may be blown away by the wind before it is eaten. Leaves may die and fall to the ground before they are eaten.
- Not all the grass is available to be eaten. The roots, for example, are underground where few animals will find and eat them.
- Of the grass that is eaten, much is indigestible inside the alimentary canals of the herbivores. Cellulose and lignin are difficult to digest and may simply pass out in the faeces rather than being absorbed into the herbivores' bodies.
- The grass plants require energy themselves, which they obtain by respiration. This breaks down organic molecules to carbon dioxide and water, and the energy is eventually lost as heat, so is no longer available to herbivores.

The diagram shows the quantities of energy transferred between organisms in a food chain in a salt marsh. The figures are in kJ m^{-2} y^{-1}. Only three trophic levels are shown.

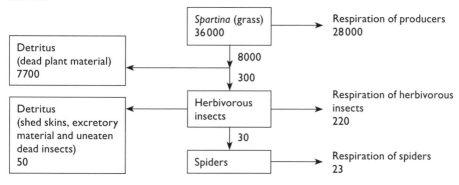

Energy transfer in a salt marsh food chain

We can use this diagram to calculate the efficiency of energy transfer between the primary consumers (herbivorous insects) and the secondary consumers (spiders).

$$efficiency = \frac{kJ \text{ of energy transferred to secondary consumers}}{kJ \text{ of energy transferred to primary consumers}} \times 100$$

$$= \frac{30}{300} \times 100 = 10\%$$

The nitrogen cycle

Living organisms need nitrogen because nitrogen atoms are an essential part of proteins, nucleic acids and ATP.

The air contains about 78% nitrogen gas. However, this is in the form of nitrogen molecules, in which two nitrogen atoms are held together by a very strong triple covalent bond. This is very unreactive. Nitrogen molecules freely diffuse in and out of the bodies of living organisms, but take no part in the metabolic reactions inside their cells.

Nitrogen fixation

For nitrogen to become involved in metabolic reactions, it must first be converted to a different form by combining with oxygen or hydrogen. This process is called **nitrogen fixation**. It can be done by:

- **lightning**, which provides very high temperatures that can cause nitrogen and oxygen molecules in the air to combine to form nitrogen oxides; these can then be washed to the ground in rain.
- **industrial processes** in which nitrogen is combined with hydrogen to produce ammonia, NH_3; this is then used to manufacture fertilisers such as ammonium nitrate.
- **nitrogen-fixing bacteria**, which use the enzyme nitrogenase to combine nitrogen and hydrogen to produce ammonium ions. Some of these bacteria live free in the soil, lakes or oceans. Others, for example *Rhizobium*, live symbiotically in root nodules in several different species of plants, particularly legumes such as peas and beans.

Formation of amino acids

Plants are able to take nitrate ions, NO_3^-, or ammonium ions, NH_4^+, from the soil into their root hairs. This may be done by diffusion or active transport. These ions can be combined with carbohydrates to produce amino acids.

Consumers obtain their nitrogen by eating proteins and other nitrogen-containing organic compounds that were originally synthesised by plants.

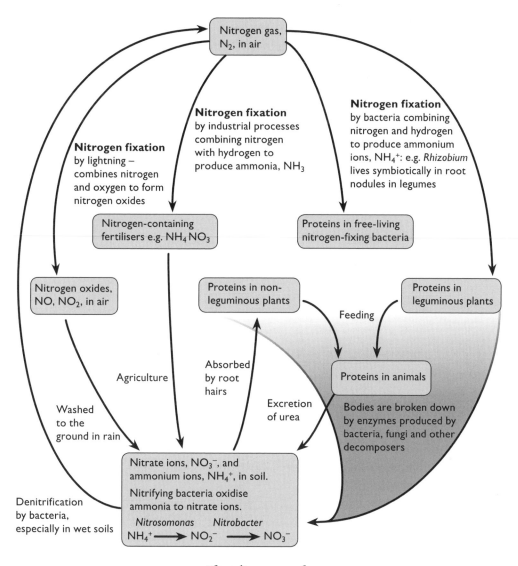

The nitrogen cycle

Decay and ammonification

Animals excrete nitrogen-containing compounds such as ammonia and urea. When they die, protein molecules in their bodies are broken down by enzymes produced by bacteria, fungi and other decomposer organisms. These processes add ammonia and ammonium ions to the soil.

Nitrification

Nitrifying bacteria oxidise ammonia to nitrate ions. This is done in two stages:
- *Nitrosomonas* oxidises ammonium ions to nitrite ions, NO_2^-;
- *Nitrobacter* oxidises nitrite ions to nitrate ions, NO_3^-.

The nitrate ions can then be taken up by plant roots.

Denitrification

Several different types of bacteria get their energy by converting nitrate ions to nitrogen gas. This process is called denitrification, and it returns nitrogen gas to the atmosphere.

AS
Experimental
Skills &
Investigations

AS Experimental skills and investigations

Almost one quarter of the total marks for your AS examination are for experimental skills and investigations. These are assessed on Paper 3, which is a practical examination.

There is a total of 40 marks available on this Paper. Although the questions are different on each Paper 3, the number of marks assigned to each skill is always the same. This is shown in the table below.

Skill	Total marks	Breakdown of marks	
Manipulation, measurement and observation, MMO	16	Successfully collecting data and observations	8 marks
		Making decisions about measurements or observations	8 marks
Presentation of data and observations, PDO	12	Recording data and observations	4 marks
		Displaying calculations and reasoning	2 marks
		Data layout	6 marks
Analysis, conclusions and evaluation, ACE	12	Interpreting data or observations and identifying sources of error	6 marks
		Drawing conclusions	3 marks
		Suggesting improvements	3 marks

The syllabus explains each of these skills in detail, and it is important that you read the appropriate pages in the syllabus so that you know what each skill is, and what you will be tested on.

The next few pages explain what you can do to make sure you get as many marks as possible for each of these skills. They give you guidance in how you can build up your skills as you do practical work during your course, and also how to do well in the examination itself. They are not arranged in the same order as in the syllabus, or in the table above. Instead, they have been arranged by the kind of task you will be asked to do, either in practical work during your biology course or in the examination.

There is a great deal of information for you to take in, and skills for you to develop. The only way to do this really successfully is to do lots of practical work, and gradually build up your skills bit by bit. Don't worry if you don't get everything right first time. Just take note of what you can do to improve next time — you will steadily get better and better.

The examination questions

There are usually two questions on Paper 3. The examiners will take care to set questions that are **not exactly the same** as any you have done before. It is possible that there could be three shorter questions instead of two longer ones, so do not be surprised if that happens.

It is very important that you do exactly what the question asks you to do. Candidates often lose marks by doing something they have already practised, rather than doing what the question actually requires.

Question 1

This is likely to be what is sometimes called a 'wet practical'. For example, it could be:
- an investigation into the activity of an enzyme
- an osmosis experiment
- tests for biological molecules

This question will often ask you to investigate the effect of one factor on another — for example, the effect of enzyme concentration on rate of reaction, or the effect of leaf area on the rate of transpiration.

Question 2

This question is likely to involve making drawings from a specimen. This could be a real specimen, or it could be a photograph. You may be asked to use a microscope, a stage micrometer and eyepiece graticule, or images of them, to work out the magnification or size of the specimen.

The two questions are designed to take up approximately equal amounts of your time. You should therefore aim to spend about 1 hour on each question.

> **Tips** During your course:
> - Every time you do a practical during your AS course, time yourself. Are you working quickly enough? You will probably find that you are very slow to begin with, but as the course progresses try to work a little faster as your confidence improves.
>
> In the exam:
> - Do exactly what the question asks you to do. This is unlikely to be exactly the same as anything you have done before.
> - Leave yourself enough time to do each question, spending an appropriate number of minutes on each one.

AS experimental skills & investigations

How to get high marks in Paper 3

Variables

Many of the experiments that you will do during your AS course, and usually Question 1 in the examination paper, will investigate the effect of one factor on another. These factors are called **variables**.

Types of variables

The factor that you change or select is called the **independent variable**. The factor that is affected (and that you measure when you collect your results) is the **dependent variable**. The table shows some examples.

Some examples of investigations

Investigation		Independent variable	Dependent variable
1	Investigation into the effect of temperature on the rate of breakdown of hydrogen peroxide by catalase	temperature	volume of oxygen produced per minute
2	Investigating the effect of immersion in solutions of different sucrose concentration on the change in length of potato strips	sucrose concentration	change in length of potato strip
3	Testing the hypothesis: the density of stomata on the lower surface of a leaf is greater than the density on the upper surface	upper or lower surface of the leaf	number of stomata per cm^3
4	Investigation into the effect of leaf area on transpiration rate	total area of leaves	rate of movement of meniscus

We will keep referring back to these four examples in the next few pages, so you might like to put a marker on this page so you can easily flip back to look at the table as you read.

If you are investigating the effect of one variable on another, then you need to be sure that there are no other variables that might be affecting the results. It is important to identify these and — if possible — keep them constant. These are sometimes called **control variables**.

Making decisions about the independent variable

You may have to make your own decisions about the range and interval of the independent variable.

Let's think about **Investigation 1** in the table above — **investigating the effect of temperature on the rate of breakdown of hydrogen peroxide by catalase.**

The independent variable is the temperature. First, decide on the **range** of temperatures you will use. The range is the spread between the highest and lowest value. This will be affected by:

- the apparatus you have available to you, which will determine the possible range of temperatures you can produce. In this case, you will probably be using a water bath. If you are lucky, you may have a thermostatically controlled water bath, but in the exam you will probably have to use a beaker of water whose temperature you can control by adding ice or by heating it.

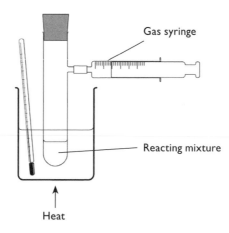

Changing the independent variable

- your knowledge about the range of temperature over which the rate of activity of the enzyme is likely to be affected. Even if you could manage it, there wouldn't be much point in trying temperatures as low as −50 °C or as high as 200 °C. However, you probably know that various enzymes can have optimum temperatures anywhere between 20 °C and 80 °C, so you should include these values in the range.

Next, decide on the **intervals** that you will use. The interval is the distance between the values that you choose. This will be affected by:

- the number of different values you can fit in within your chosen range, and how much time you have available to you. For example, you might ideally like to use intervals of 5 °C, so that you set up water baths at 0 °C, 5 °C and so on, all the way up to 80 °C. But obviously that would not be sensible if you only have five water baths, or if you only have 1 hour to do the experiment.
- the number of results you need to obtain. You are going to be looking for any pattern in the relationship between the independent variable (temperature) and the dependent variable (rate of reaction). You will need at least 5 readings to see any pattern. There is really no point trying to draw a graph if there will be fewer than 5 points on it. So, if your range of temperatures is 0 °C to 80 °C, you could use intervals of 20 °C. This would give you 5 readings — 0, 20, 40, 60 and 80 °C.

Producing different concentrations of a solution

In **Investigation 2, Investigating the effect of immersion in solutions of different sucrose concentration on the change in mass of potato strips**, the independent variable is the concentration of a solution. You may be given a sucrose solution of a particular concentration, and then be asked to produce a suitable range of concentrations to carry out the experiment.

The *range* you should use will usually be from 0 (distilled water) up to the concentration of the solution you have been given (because obviously you can't easily make that into a more concentrated solution).

The *intervals* you use could be either:

- all the same distance apart, for example concentrations of 0.8, 0.6, 0.4 and 0.2 mol dm^{-3} (and, of course, 0.0 mol dm^{-3})
- produced by using serial dilutions to make concentrations of 0.1, 0.01 and 0.001 mol dm^{-3} (and, of course, 0.0 mol dm^{-3})

Producing a range of solutions of different concentrations from one given concentration

Let's say you need 10 cm^3 of sucrose solution of each concentration.

How to produce a range with equal intervals

Take a particular volume of your original solution and place it in a clean tube. Add distilled water to make it up to 10 cm^3.

Then do the same again, using a different volume of the original solution.

The table below gives some examples.

Producing a range of concentrations with equal intervals from a 1.0 mol dm^{-3} solution

Volume taken of original 1 mol dm^{-3} solution/cm^3	Volume of distilled water added/cm^3	Concentration of solution produced/ mol dm^{-3}
10	0	1.0
8	2	0.8
6	4	0.6
4	6	0.4
2	8	0.2
0	10	0.0

How to produce a range using serial dilutions

You could be asked to make up a series of solutions in which each one has a concentration that is one tenth of the previous one.

Take 1 cm^3 of your original solution and place it in a clean tube.

Add distilled water to make it up to 10 cm^3.

Now mix this new solution really well, and then take $1\,cm^3$ of it. Put this into a clean tube and make it up to $10\,cm^3$.

Keep doing this, each time taking $1\,cm^3$ from the new solution.

The table below summarises this.

Producing a range of concentrations using serial dilution of a $1.0\,mol\,dm^{-3}$ solution

Solution used/ $mol\,dm^{-3}$	Volume taken of solution/cm^3	Volume of distilled water added/cm^3	Concentration of solution produced/ $mol\,dm^{-3}$
1.0	10	0	1.0
1.0	1	9	0.1
0.1	1	9	0.01
0.01	1	9	0.001
0.001	1	9	0.0001
0.0001	1	9	0.00001

You could also be asked to make up solutions where each is one half of the concentration of the previous solution.

Continuous and discontinuous variables

In Investigation 1, the independent variable (temperature) is **continuous**. This means that we can choose any value within the range we have decided to use. This is also true for Investigation 2, where we can choose any value of concentration within the range we have decided to use.

Sometimes, however, the independent variable is **discontinuous**. This means that there is only a limited number of possible values. For example, in **Investigation 3, Testing the hypothesis: the density of stomata on the lower surface of a leaf is greater than the density on the upper surface**, the independent variable has only two possible 'values' — either the upper surface of the leaf, or the lower surface of the leaf. So you don't have any choice about the range or intervals at all!

Tips During your course:
- Every time you do an experiment, identify and write down the independent variable and the dependent variable.
- Every time you do an experiment, think about the *range* and the *intervals* of the values you are using for the independent variable. For your own benefit, write down what the range is and what the intervals are, just to help you to think about them.
- Learn how to make up dilutions from a solution of a given concentration, and practise doing this until you feel really confident about it.

In the exam:

- Read the question carefully, then identify the independent variable and the dependent variable (even if the question does not ask you to do this).
- Next, decide if the independent variable is continuous or discontinuous (see above).
- If it is continuous, read the question carefully to see if you have been told the range and intervals to use, or if you are being asked to decide these for yourself.

Controlling the control variables

In your experiment, it is important to try to make sure that the only variable that could be affecting the dependent variable is the independent variable that you are investigating. If you think there are any other variables that might affect it, then you must try to keep these constant.

Look back at the Table on page 100.

In Investigation 1, the important control variables would be the concentration and volume of the enzyme solution and the concentration and volume of the hydrogen peroxide solution. Changes in any of these would have a direct effect on the rate of reaction.

In Investigation 2, the important control variables would be the dimensions of the potato strips and the potato tuber from which they came. You also need to think about time, but here the important thing is that the strips are left in the solution for long enough for equilibrium to be reached — after that, it doesn't really matter if one is left for slightly longer than another. You also need to be sure that all the strips are completely immersed in the solution, although the actual volume of the solution doesn't matter. Temperature, too, won't affect the final result, but it could affect the speed at which equilibrium is reached — if you leave the strips for long enough, then it does not really matter if the temperature varies.

In biology, we often want to do experiments where it is not possible to control all the variables. For example, we might want to investigate the effect of body mass index on heart rate when at rest. There are all sorts of other variables that might affect resting heart rate, such as gender, age, fitness, when a person last ate and so on. In this case, we just have to do the best we can, for example, by limiting our survey to males between the ages of 20 and 25. If we can collect results from a large *random* sample among this group of males, then we can hope that at least we will be able to see if there appears to be a relationship between our independent and dependent variables.

Tips During your course:

- Every time you do an experiment, think about which variables you have been told to control, or make your own decision about which ones are important to control. Get to know the standard ways of controlling variables such as temperature (use water baths), pH (use buffer solutions) and other variables.

In the exam:

- If you are not told what variables to control, then think about these carefully before deciding what you will control and how you will do it.

When to measure the dependent variable

In many experiments you will need to decide when, and how often, you should take a reading, observation or measurement of the dependent variable.

- With some investigations, you will need to leave things long enough for whatever is happening to finish happening. This would be important in Investigation 2, where you would need to leave the potato strips in the sucrose solutions for long enough for equilibrium to be reached.
- With some investigations, you may need to begin taking readings straight away. This would be important in Investigation 1, where you should begin measuring the volume of oxygen released each minute from time 0, which is the moment that the enzyme and its substrate are mixed.
- With some investigations, you may need to allow time for a process to settle down to a steady rate before you begin to take readings. This would be important in Investigation 4, where you would be measuring the rate of transpiration in a particular set of conditions.

> **Tips** During your course:
> - Every time you do an experiment where time is involved, think about why you should start timing from a particular moment, and when and why you should take readings.
>
> In the exam:
> - Think carefully about whether or not time is important. If you think it is, then decide when you will start taking readings, and how often you will take them. Remember that if you are going to use them to draw a graph, you will need at least 5 points to plot.

Taking measurements

You will often be asked to take measurements or readings. In biology, these are most likely to be length, mass, time, temperature or volume. You could be taking readings from a linear scale (for example, reading temperature on a thermometer, reading volume on a pipette, or reading length on a potometer tube). You could be reading values on a digital display, for example reading mass on a top pan balance or time on a digital timer.

There are some special terms that are used to describe measurements, and the amount of trust you can put into them. It's easiest if we think about them in terms of a particular experiment, so let's concentrate on Investigation 1. Look back at page 100 to remind yourself what is being measured.

Validity This is about whether what you are measuring is what you actually *intend* to measure. For example, in Investigation 1, does measuring the volume of oxygen in the gas syringe each minute really tell you about the rate of reaction? It is a valid method in this instance, because the volume of oxygen given off per unit time is directly related to the rate at which the reaction is taking place.

Reliability   This is how well you can trust your measurements. Reliable results are ones that are repeatable. This could be affected by various factors, such as whether you are able to take a reading at the precise time you intended to.

Accuracy   An accurate reading is a true reading. For your readings of volume to be accurate, then the gas syringe must have been calibrated correctly, so that when it says the volume of gas is $8.8\,cm^3$, then there really is exactly $8.8\,cm^3$ of gas in there.

Precision   If you were able to put exactly $8.8\,cm^3$ of gas into your gas syringe, and it read $8.8\,cm^3$ every time, then your readings have a high degree of precision. If, however, the syringe didn't always read the same value (so there was variation in its readings, even though the actual volume of gas was exactly the same), then your measurements are less precise.

Resolution   You probably already know this term, because we use it in microscopy to tell us the degree of detail that we can see. The smaller the detail, the higher the resolution. It means very much the same thing with a measuring instrument — the smaller the division on the scale of the measuring instrument, the higher its resolution. So, for example, a $10\,cm^3$ gas syringe marked off every $0.5\,cm^3$ has a higher resolution than a $20\,cm^3$ gas syringe marked off every $1\,cm^3$. If you get a choice, then go for the instrument with the highest resolution to make your measurements — so long as it can cover the range that you need.

Uncertainty in measurements— estimating errors

Whenever you take a reading or make a measurement, there will be some uncertainty that the value is absolutely correct. These uncertainties are **experimental errors**. Every experiment, no matter how well it has been designed, no matter how carefully it has been carried out and no matter how precise and accurate the measuring instruments, has this type of error.

You may be asked to estimate the size of the errors in your measurements. **This is nothing to do with how well you have made the measurements** — the examiners don't want to know about 'mistakes' that you might have made, such as misreading a scale or taking a reading at the wrong time. It is all about the inbuilt limitations in your measuring device and its scale.

- In general, **the size of the error is half the value of the smallest division on the scale**. For example, if you have a thermometer that is marked off in values of $1\,°C$, then every reading that you take could be out by $0.5\,°C$. You can show this by writing: $21.5\,°C \pm 0.5\,°C$.
- If your recorded result involves measuring *two* values — for example, if you have measured a starting temperature and then another temperature at the end, and have calculated the rise in temperature — then this error could have occurred for both readings. **The total error is therefore the sum of the errors for each reading**. Your final value for the change of temperature you have measured would then be written: $18.0\,°C \pm 1.0\,°C$.

> **Tip**   Every time you take a reading or make a measurement, get into the habit of working out and writing down the error (uncertainty) in each reading.

Recording measurements and other data

You will often need to construct a table in which to record your measurements, readings and other observations.

It is always best to design and construct your results table *before* you begin your experiment, so that you can write your readings directly into it as you take them.

Let's think about Investigation 2 again. You've made your decisions about the range and intervals of the independent variable (concentration of solution) — you've decided to use six concentrations ranging from $0.0\,mol\,dm^{-3}$ to $1.0\,mol\,dm^{-3}$. Your dependent variable is the change in length of the potato strips, and you are going to find this by measuring the initial length and final length of each strip.

These are the things you need to think about when designing your results table:

- The **independent variable** should be in the first column.
- The **readings** you take are in the next columns.
- Sometimes, these readings actually *are* your dependent variable. In this experiment, however, you are going to have to use these readings to *calculate* your dependent variable, which is the change in length of the strips. So you need to have another column for this. This comes at the end of the table. In fact, you really need to work out the *percentage* change in length of the strips, as this will allow for the inevitable variability in the initial lengths of the strips.

The table could look like this:

Results table for Investigation 2

Concentration of sucrose solution/ $mol\,dm^{-3}$	Initial length of potato strip/mm	Final length of potato strip/ mm	Change in length of potato strip/ mm	Percentage change in length of potato strip
0				
0.2				
0.4				
0.6				
0.8				
1.0				

Notice:

- The table has been clearly drawn, with lines separating all the different rows and columns. Always use a pencil and ruler to draw a results table.
- Each column is fully headed, including the unit in which that quantity is going to be measured. The unit is preceded by a slash /. You can use brackets instead — concentration of sucrose solution $(mol\,dm^{-3})$.
- The slash always means the same thing. It would be completely wrong to write: concentration of sucrose solution/mol/dm^3 as the heading of the first column. That would be really confusing. If you are not happy using negative indices like

dm^{-3}, you can always write 'per' instead. So it would be fine to write: concentration of sucrose solution/mol per dm^3.

- The columns are all in a sensible order. The first one is the independent variable, so you can write these values in straight away, as you have already decided what they will be. The next thing you will measure is the initial length of the strip, then the final length. Then you will calculate the change in length, and finally you will calculate the percentage change in length.

So now you are ready to do your experiment and collect your results. Here is what your table might look like.

Completed results table for Investigation 2

Concentration of sucrose solution/ $mol\,dm^{-3}$	Initial length of potato strip/mm	Final length of potato strip/ mm	Change in length of potato strip/ mm	Percentage change in length of potato strip
0	49.5	52.5	+3.0	+6.1
0.2	50.0	52.0	+2.0	+4.0
0.4	50.5	51.5	+1.0	+2.0
0.6	50.0	50.5	+0.5	+1.0
0.8	49.0	48.0	−1.0	−2.0
1.0	49.5	47.5	−2.0	−4.0

Notice:
- All the measurements in the second two columns were made to the nearest 0.5 mm. This is because the smallest graduation on the scale on the ruler was 1 mm. So it was possible to estimate the length to the nearest 0.5 mm. (Have a look at the scale on your ruler, and you will see that this is sensible.) Even if you decide that a length is exactly 50 mm, you must write in the next decimal place for consistency, so you would write 50.0.
- The values in the 'change in length' column each show whether they were an increase or a decrease.
- The percentage change in length is calculated like this:

$$\frac{\text{change in length}}{\text{initial length}} \times 100$$

(Do make sure you remember to take a calculator into the exam with you.)
- Each percentage change in length has been rounded up to one decimal place, for consistency with the change in length. For example, the calculation in the first row gives 6.0606, which you should round up to 6.1. The calculation in the sixth row gives 4.0404, which rounds up to 4.0.

Repeats
It is a good idea to do **repeats**. This means that, instead of getting just one reading for each value of your independent variable, you collect two or three. You can then calculate a **mean value**, which is more likely to be a 'true' value than any of the individual ones.

Let's say that you did this for the potato strip experiment. You could have used two potato strips for each sucrose concentration, then calculated the percentage change in length for each one, then finally calculated a mean percentage change.

This means adding some extra rows and an extra column to the results table, like this:

Completed results table (with repeats) for Investigation 2

Concentration of sucrose solution/ $mol\,dm^{-3}$	Initial length of potato strip/mm	Final length of potato strip/mm	Change in length of potato strip/mm	Percentage change in length of potato strip	Mean percentage change in length of potato strip
0	49.5	52.0	+2.5	+5.1	+4.6
	49.0	51.0	+2.0	+4.1	
0.2	50.0	52.0	+2.0	+4.0	+4.0
	50.5	52.5	+2.0	+4.0	
0.4	50.5	51.5	+1.0	+2.0	+2.5
	49.5	51.0	+1.5	+3.0	
0.6	50.0	50.5	+0.5	+1.0	+0.5
	51.0	51.0	0.0	0.0	
0.8	49.0	48.0	−1.0	−2.0	−2.0
	50.5	49.5	−1.0	−2.0	
1.0	49.5	48.0	−1.5	−3.0	−3.0
	50.5	49.0	−1.5	−3.0	

Notice:
- All of this information has been put in a single results table. This makes it much easier for someone to read and find all the information they need.
- The numbers in the final column have again been rounded up to one decimal place.

Qualitative observations
The results table for the potato strip experiment contains numerical values — they are **quantitative**. Sometimes, though, you may want to write descriptions in your results table, for example a colour that you observed. These are **qualitative** observations. If you are recording colours, write down the actual colour — do not just write 'no change'.

Use simple language that everyone can easily understand. Avoid using terms that are difficult for the examiner to interpret, such as 'yellowish-green'. Think about what is important — perhaps it is that *this* tube is a darker or lighter green than *that* tube. Using simple language such as 'dark green' or 'a lighter green than tube 1' is fine.

Graphs and other ways of displaying data

When you have collected your data and completed your results table, you will generally want to display the data so that anyone looking at them can see any patterns.

AS experimental skills & investigations

Line graphs

Line graphs are used when both the independent variable and the dependent variable are continuous (see page 103). This is the case for the potato strip data on page 109. The graph can help you to decide if there is a relationship between the independent variable and the dependent variable. This is what a line graph of these data might look like.

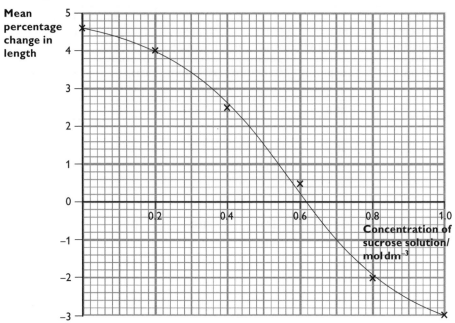

Graph of the results of Investigation 2

Notice:

- The independent variable goes on the x-axis, and the dependent variable goes on the y-axis.
- Each axis is fully labelled with units. You can just copy the headings from the appropriate columns of your results table.
- The scales on each axis should start at or just below your lowest reading, and go up to or just above your highest reading. Think carefully about whether you need to begin at 0 on either of the axes, or if there is no real reason to do this.
- The scales use as much of the width and height of the graph paper as possible. If you are given a graph grid on the exam paper, the examiners will have worked out a sensible size for it, so you should find your scales will fit comfortably. The greater the width and height you use, the easier it is to see any patterns in your data once you have plotted them.
- The scale on each axis goes up in regular steps. Choose something sensible, such as 1s, 2s, 5s or 10s. If you choose anything else, such as 3s, it is practically impossible to read off any intermediate values. Imagine trying to decide where 7.1 is on a scale going up in 3s...

- Each point is plotted very carefully with a neat cross. Don't use just a dot, as this may not be visible once you've drawn the line. You could, though, use a dot with a circle round it.
- A smooth best-fit line has been drawn. This is what biologists do when they have good reason to believe there is a smooth relationship between the independent and dependent variables. You know that your individual points may be a bit off this line (and the fact that the two repeats for each concentration were not always the same strongly supports this view), so you can actually have more faith in there being a smooth relationship than you do in your plots for each point.

Sometimes in biology (it doesn't often happen in physics or chemistry!) you might have more trust in your individual points than in any possible smooth relationship between them. If that is the case, then you do not draw a best-fit curve. Instead, join the points with a very carefully drawn straight line, like this:

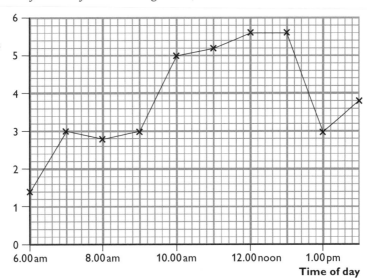

***Graph where we are not sure of the pattern in the relationship between the
independent and dependent variables***

Tips During your course:
- Get plenty of practice in drawing graphs, so that it becomes second nature always to choose the correct axes, to label them fully and to choose appropriate scales.

In the exam:
- Take time to draw your graph axes and scales — you may need to try out two or even three different scales before finding the best one.
- Take time to plot the points — and then go back and check them.
- Use a sharp HB pencil to draw the line, taking great care to touch the centre of each cross if you are joining points with straight lines. If you go wrong, rub the line out completely before starting again.
- If you need to draw two lines on your graph, make sure you label each one clearly.

AS experimental skills & investigations

You may be asked to read off an intermediate value from the graph you have drawn. It is always a good idea to use a ruler to do this — place it vertically to read a value on the x-axis, and horizontally to do the same on the y-axis. You can draw in faint vertical and horizontal pencil lines along the ruler. This will help you to read the value accurately.

You could also be asked to work out the gradient of a line on a graph. This is explained on page 40.

> **Tips** During your course:
> • Make sure you know how to read off an intermediate value from a graph accurately, and how to calculate a gradient.
>
> In the exam:
> • Take time over finding intermediate values on a graph — if you rush it is very easy to read off a value that is not quite correct.

Histograms

A histogram is a graph where there is a continuous variable on the x-axis, and a frequency on the y-axis. For example, you might have measured the length of 20 leaves taken from a tree. You could plot the data like this:

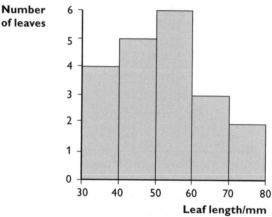

A frequency histogram

Notice:
• The numbers on the x-axis scale are written *on* the lines. The first bar therefore includes all the leaves with a length between 30 and 39 mm. The next bar includes all the leaves with a length between 40 and 49 mm, and so on.
• The bars are all the same width.
• The bars are all touching — this is important, because the x-axis scale is continuous, without any gaps in it.

Bar charts

A bar chart is a graph where the independent variable is made up of a number of different, discrete categories and the dependent variable is continuous. For example, the independent variable could be type of fruit juice, and the dependent variable could be the concentration of glucose in the juice.

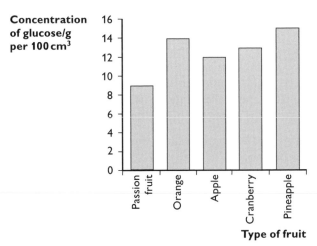

Bar chart showing concentration of glucose in different types of fruit juice

Notice:
- The x-axis has an overall heading (type of fruit), and then each bar also has its own heading (orange, apple and so on on).
- The y-axis has a normal scale just as you would use on a line graph.
- The bars are all the same width.
- The bars do not touch.

Drawing conclusions and interpreting your data

Once you have collected, tabulated and displayed your results, you can use them to draw a conclusion.

When you are thinking about a conclusion, look right back to the start of your experiment where you were told (or you decided) what you were to investigate. For example:
- in Investigation 1, Investigating the effect of temperature on the rate of breakdown of hydrogen peroxide by catalase, your conclusion should provide an answer to this question.
- in Investigation 2, Investigating the effect of immersion in solutions of different sucrose concentration on the change in length of potato strips, your conclusion should state the relationship between the concentration of sucrose solution and the change in length of the potato strips.
- in Investigation 4, Testing the hypothesis: the density of stomata on the lower surface of a leaf is greater than the density on the upper surface, your conclusion should say whether your results support or disprove this hypothesis.

Explaining your reasoning

There will often be marks for explaining how you have reached your conclusion. Your reasoning should refer clearly to your results. For example, your conclusion to Investigation 2 (whose results are shown in the table on page 109) might be:

> A sucrose solution with a concentration of $0.6\,\text{mol}\,\text{dm}^{-3}$ and below caused an increase in length of the potato strips. A sucrose solution with a concentration of $0.8\,\text{mol}\,\text{dm}^{-3}$ and above caused a decrease in length of the potato strips. From the graph, the solution that I would expect to cause no change in length of the strips would be $0.62\,\text{mol}\,\text{dm}^{-3}$.

> The strips gained in length because they took up water, which was because the water potential of the sucrose solution was greater than the water potential in the potato cells. This therefore means that the water potential inside the potato cells was the same as the water potential of a $0.62\,\text{mol}\,\text{dm}^{-3}$ sucrose solution.

Showing your working, and significant figures

You may be asked to carry out a calculation and to show your working. There will be marks for doing this. If you do not show your working clearly, then you will not get full marks, even if your answer is absolutely correct.

For example, imagine you have measured four lengths as 46 mm, 53 mm, 52 mm and 48 mm. You are asked to calculate the mean and to show your working. You should write this down properly as:

$$\text{mean length} = \frac{46 + 53 + 52 + 48}{4}\ \text{mm} = 50\,\text{mm}$$

You've already seen, on page 108, that the final answer to a calculation should have the same number of significant figures as the original numbers you were working from. If you do the calculation above, you'll find the answer you get is 49.75. But the original measurements were only to two significant figures (a whole number of mm) so that is how you should give the final answer to your calculation. You must round the answer up or down to give the same number of significant figures as the original values from which you are working.

There's another example of showing your working on page 119.

> **Tips** During your course:
> - Get into the habit of describing the main steps in your reasoning when drawing a conclusion, and using evidence from the results to support it.
> - Get into the habit of taking time to set out all your calculations very clearly, showing each step in the process.
> - Get into the habit of giving final numerical answers to calculations to the same number of significant figures as the readings you took, or the values you were given.
>
> In the exam:
> - Even if you feel rushed, take time to write down the steps in calculations and reasoning fully.

Identifying sources of error

It's worth repeating that it is very important to understand the difference between experimental errors and 'mistakes'. A mistake is something that you do incorrectly, such as misreading the scale on a thermometer, or taking a reading at the wrong time, or not emptying a graduated pipette fully. Do **not** refer to these types of mistake when you are asked to comment on experimental errors.

You've already seen, on page 106, that every measuring instrument has its own built-in degree of uncertainty in the values you read from it. You may remember that, in general, the size of the error is half the value of the smallest division on the scale.

Errors can also occur if there were uncontrolled variables affecting your results. For example, if you were doing an investigation into the effect of leaf area on the rate of transpiration, and the temperature in the laboratory increased while you were doing your experiment, then you can't be sure that all the differences in rate of transpiration were entirely due to differences in leaf area.

Systematic and random errors

Systematic errors are ones that are the same throughout your investigation, such as intrinsic errors in the measuring instruments you were using.

Random errors are ones that can differ throughout your investigation. For example, you might be doing an osmosis investigation using potato strips taken from different parts of a potato, where perhaps the cells in some parts had a higher water potential than in others. Or perhaps the temperature in the room was fluctuating up and down.

Spotting the important sources of error

You should be able to distinguish between significant errors and insignificant ones. For example, a change in room temperature could have a significant effect on the rate of transpiration (Investigation 4) but it would not have any effect at all on the number of stomata on the upper and lower surface of a leaf (Investigation 3).

Another thing to consider is how well a variable has been controlled. If you were doing an enzyme investigation using a water bath to control temperature, then you should try to be realistic in estimating how much the temperature might have varied by. If you were using a high-quality, electronically controlled water bath, then it probably did not vary much, but if you were using a beaker and Bunsen burner then it is likely that temperature variations could indeed be significant.

> **Tips** During your course:
> - Every time you do an investigation, work out and write down the uncertainty in all the types of measurement that you make.
> - Every time you do an investigation, think carefully about any errors that may be due to lack of control of variables — which ones might genuinely be significant?
>
> In the exam:
> - If you are asked about an investigation that seems familiar, it is tempting just to try to recall what the main errors were in the investigation that you did before.

This is not a good idea, because the investigation in the exam may not be quite the same. Always think about the actual investigation in the examination question, and *think through* what the significant sources of error are.

Suggesting improvements

You may be asked to suggest how the investigation you have just done, or an investigation that has been described, could be improved. Your improvements should be aimed at getting more valid or reliable results to the question that the investigation was trying to answer — do not suggest improvements that would mean you would now be trying to answer a different question. For example, if you were doing an investigation to investigate the effect of *leaf area* on the rate of transpiration, don't suggest doing something to find out the effect of the *wind speed* on the rate of transpiration.

The improvements you suggest could include controlling certain variables that were not controlled, or controlling them more effectively. For example, you may suggest that the investigation could be improved by controlling temperature. To earn a mark, you must also say *how* you would control it, for example by placing sets of test-tubes in a thermostatically controlled water bath.

You could also suggest using better methods of measurement. For example, you might suggest using a colorimeter to measure depth of colour, rather than using your eyes and a colour scale.

It is almost always a good idea to do several repeats in your investigation and then calculate a mean of your results. For example, if you are measuring the effect of light intensity on the rate of transpiration, then you could take three sets of readings for the volume of water taken up by your leafy shoot in one minute at a particular light intensity. The mean of these results is more likely to give you the true value of the rate of transpiration than any one individual result.

> **Tips** During your course:
> - If time allows, try to do at least two (and possibly three) repeats when you do an investigation.
> - As you do an investigation, be thinking all the time about how reliable or accurate your measurements and readings are. Think about what you would like to be able to do to improve their reliability or accuracy.
>
> In the exam:
> - Be very precise in suggesting how you could improve the investigation — for example, don't just say you would control a particular variable, but say *how* you would control it.

Drawings

One of the questions in the exam is likely to involve drawing a specimen on a slide, using a microscope, or drawing from a photomicrograph (a photograph taking through a microscope).

Making decisions about what to draw

You might have to decide which part of a micrograph to draw. For example, there might be a micrograph of a leaf epidermis, and you are asked to draw two guard cells and four epidermal cells. It is really important that you do exactly as you are asked and choose an appropriate part of the micrograph.

Producing a good drawing

It is very important that you draw what you can see, not what you think you ought to see. For example, during your AS course you may have drawn a TS of a stem where the vascular bundles were arranged in a particular way, or were a particular shape. In the exam, you could be asked to draw a completely different type of vascular bundle that you have never seen before. Look very carefully and draw what you can see.

Your drawing should:

- be large and drawn using a sharp pencil (preferably HB, which can be easily erased if necessary) with no shading, using single, clear lines;
- show the structure or structures in the correct proportions. The examiners will check that the overall shape and proportions of your drawing match those of the specimen. Don't worry — you don't need to be a wonderful artist — a simple, clear drawing is all that is required;
- show only the outlines of tissues if you are asked to draw a low power plan (LPP). A LPP should **not** show any individual cells. However, if you are using a microscope, you may need to go up to high power to check exactly where the edges of the tissues are.

You may be asked to label your drawing. In that case:

- use a pencil to draw label lines to the appropriate structure using a ruler, ensuring that the end of the label line actually touches the structure you are labelling;
- make sure that none of your label lines cross each other;
- write the actual labels horizontally;
- write the actual labels outside the drawing itself.

> **Tips** During your course:
> - Make sure you are familiar with the appearance of all of the structures listed in the syllabus that you could be asked about on the practical paper. You need to know the names and distribution of the tissues. Look in particular at the learning outcomes marked with [PA] at the beginning.
> - Practise drawing specimens from micrographs, getting used to using your own eyes to see what is really there, rather than what you think ought to be there;
> - Practise using an eyepiece graticule to help you work out the relative proportions of different parts that you are drawing.
> - Take every opportunity to practise drawing specimens from micrographs or microscope slides, and either mark them yourself using a CIE-style mark scheme, or get your teacher to mark them for you. Find out what you need to do to improve, and keep working at it until you feel really confident.

AS experimental skills & investigations

In the exam:
- Take one or two sharp HB pencils, a pencil sharpener, a clean ruler that measures in mm and a good eraser.
- Settle down and take time to get your drawing of the specimen right.
- Use your eyes first, then your memory.

Calculating magnification or size

The use of a stage micrometer and eyepiece graticule is described on pages 16–17. You might be asked to do this on Paper 3.

You could also be given the magnification of an image, and asked to calculate the real size of something in the image. Below is an example of the kind of thing you might be asked to do.

This micrograph shows some cells from a moss. Notice that the magnification is given.

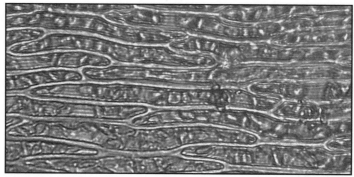

Magnification × 300

Let us say you are asked to find the mean width of a cell from the tissue in the micrograph. There are several steps you need to work through here.

First, decide *how many* cells you are going to measure. It is generally sensible to measure a randomly selected sample of 5 to 10 cells.

Next, decide *which ones* you will measure. Choose cells where you can see the edges as clearly as possible, and where you can see the whole cell. If cells are evenly distributed, it is best to measure the total width of five cells in a row. That means you have to make fewer measurements, do fewer calculations and — better still — it reduces the size of the uncertainty in your measurements. However, if cells are irregularly shaped or distributed, you should measure each one individually.

Once you have decided which five cells to measure, mark this clearly on the micrograph. It doesn't matter exactly how you do this — perhaps you could carefully use a ruler to draw a line across the five cells, beginning and ending exactly at the first edge of the first cell, and the last edge of the fifth cell.

Now measure the length of the line in mm and write it down.

Next, calculate the mean length of one cell. Show clearly how you did this.

Next, convert this length in mm to a length in μm. (Alternatively, you could do this right at the end of the calculation.)

Next, use the magnification you have been given to convert this mean length of the image to a mean real length.

Here is what your answer might look like:

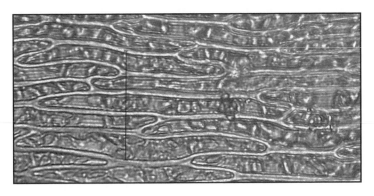

total width of 5 cells on micrograph = 29 mm

therefore mean width = $\frac{29}{5}$ mm = 6 mm

$$= 6 \times 1000 = 6000 \, \mu m$$

magnification = × 300

therefore real mean width of a cell = $\frac{6000}{300} \, \mu m = 20 \, \mu m$

Making comparisons

You may be asked to compare the appearance of two biological specimens or structures. You could be observing these using the naked eye or a lens, or using a microscope, or you could be looking at two micrographs.

The best way to set out a comparison is to use a table. It will generally have three columns, one for the feature to be compared, and then one for each of the specimens.

For example, you might be asked to observe two leaves and record the differences between them. Your table and the first three differences might look like this:

Feature	Leaf A	Leaf B
Leaf margin	Smooth	Toothed
Veins	Parallel to each other	A central vein with branches coming off it, forming a network
Shape	Length is more than twice the maximum width	Length is less than twice the maximum width

Notice:
- The table has been drawn with ruled lines separating the columns and rows.

- The descriptions of a particular feature for each specimen are opposite one another (that is, they are in the same row).
- Each description says something positive. For example, in the first row, it would not be good to write 'not toothed' for Leaf A, as that does not tell us anything positive about the leaf margin.

Note that the practical examination is likely to ask you to describe or compare observable features, **not** functions. Do not waste time describing functions when this is not asked for.

AS
Questions
&
Answers

In this section is a practice examination paper, similar to the Cambridge International AS Level Biology Paper 2. All of the questions are based on the topic areas described in the previous sections of the book.

You have 1 hour and 15 minutes to do the paper. There are 60 marks on the paper, so you can spend just over one minute per mark. If you find you are spending too long on one question, then move on to another that you can answer more quickly. If you have time at the end, then come back to the difficult one.

Some of the questions require you to recall information that you have learned. Be guided by the number of marks awarded to suggest how much detail you should give in your answer. The more marks there are, the more information you need to give.

Some of the questions require you to use your knowledge and understanding in new situations. Don't be surprised to find something completely new in a question — something you have not seen before. Just think carefully about it, and find something that you do know that will help you to answer it.

Do think carefully before you begin to write. The best answers are short and relevant — if you target your answer well, you can get a lot of marks for a very small amount of writing. Don't say the same thing several times over, or wander off into answers that have nothing to do with the question. As a general rule, there will be twice as many answer lines as marks. So you should try to answer a 3 mark question in no more than 6 lines of writing. If you are writing much more than that, you almost certainly haven't focused your answer tightly enough.

Look carefully at exactly what each question wants you to do. For example, if it asks you to 'Explain', then you need to say *how* or *why* something happens, not just *describe* what happens. Many students lose large numbers of marks by not reading the question carefully.

Following each question in the practice paper overleaf, there is an answer that might get a C or D grade, followed by an examiner's comments. Then there is an answer that might get an A or B grade, again followed by an examiner's comments. You might like to try answering the questions yourself first, before looking at these.

Notice that there are sometimes more ticks on the students' answers than the number of marks awarded. This could be because you need two correct responses for one mark (e.g. Q1 (a) (i)) or because there are more potential mark points than the total number of marks available (e.g. Q1 (a) (ii)). Even if you get four or five ticks for a three-mark question, you can't get more than the maximum three marks.

Exemplar paper
Question 1

(a) **The diagram shows a small part of a cell, as seen using an electron microscope.**

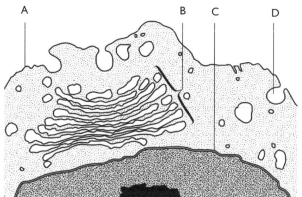

(i) **Name the parts labelled A to D.** (2 marks)

(ii) **Describe how part B is involved in the formation of extracellular enzymes.** (3 marks)

(b) **Give two reasons, other than the presence of part B, why the cell in the diagram cannot be a prokaryotic cell.** (2 marks)

Total: 7 marks

Candidate A

(a) (i)   A plasma membrane ✓
    B Golgi ✓
    C nucleus ✗
    D phagocyte ✗

 🖉 C is the nuclear envelope (or membrane), not the nucleus itself. A phagocyte is a cell — perhaps the candidate is thinking of a phagocytic vesicle. 1/2

(ii) First, the enzymes are made by protein synthesis on the ribosomes. Then they go into the endoplasmic reticulum. Then they are taken ✓ to the Golgi where they are packaged. Then they go in vesicles ✓ to the cell membrane where they are sent out by exocytosis.

 🖉 This candidate has not really thought about exactly what the question was asking, and has wasted time writing about events that take place before and after the involvement of the Golgi apparatus. There is, however, a mark for the idea that the Golgi apparatus receives proteins that have been in the RER, and another for packaging them into vesicles. 2/3

(b) It has a nucleus. ✓ And it has Golgi apparatus. ✗

    ☑ The Golgi apparatus is part B, and this has been excluded by the question. 1/2

Candidate B

(a) (i) A cell surface membrane ✓

        B Golgi apparatus ✓

        C nuclear envelope ✓

        D vesicle ✓

    ☑ All correct. 2/2

(ii) Proteins made in the RER are transported to the convex face ✓ of the Golgi apparatus in vesicles. The vesicles fuse ✓ with the Golgi and the proteins inside are modified ✓ by adding sugars to make glycoproteins. ✓ They are packaged inside membranes ✓ and sent to the cell membrane.

    ☑ All correct. 3/3

(b) If it was a prokaryotic cell it wouldn't have a nucleus ✓ and it would have a cell wall. ✓

    ☑ Correct. 2/2

Question 2

The diagram shows the bacterium _Mycobacterium tuberculosis_, which causes tuberculosis (TB).

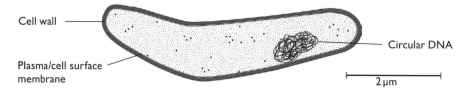

(a) _M. tuberculosis_ **is taken up by macrophages and multiplies inside them.**

Suggest how this strategy could help to protect _M. tuberculosis_ from the immune response by B cells. (3 marks)

(b) In an experiment to investigate how _M. tuberculosis_ avoids destruction by macrophages, bacteria were added to a culture of macrophages obtained from the alveoli of mice. At the same time, a quantity of small glass beads, equivalent in size to the bacteria, were added to the culture. The experiment was repeated using increasing quantities of bacteria and glass beads.

After 4 hours, the macrophages were sampled to find out how many had taken up either glass beads or bacteria. The results are shown in the graph. The x-axis shows the initial ratio of bacteria or glass beads to macrophages in the mixture.

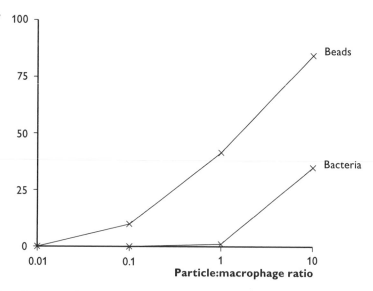

Discuss what these results suggest about the ability of macrophages to take up *M. tuberculosis*. (3 marks)

(c) When *M. tuberculosis* is present inside a phagosome of a macrophage, it secretes glycolipids that accumulate in lysosomes and prevent the lysosomes fusing with the phagosome.

Explain how this prevents the macrophage from destroying the bacterium. (3 marks)

Total: 9 marks

Candidate A

(a) It stops the B cells seeing them, so they don't make antibodies ✓ against them.

 This is not a very clear answer. B cells do not 'see', so this is not a good term to use. The 'they' in the second half of the sentence could refer to either B cells or the bacteria. 1/3

(b) The macrophages took up more glass beads than bacteria. ✓ So they are not very good at taking up the bacteria. ✓

 Just enough for two marks, although the second sentence is weak. 2/3

(c) Lysosomes contain digestive enzymes, ✓ so if they don't fuse with the phagosome the bacteria won't get digested. ✓

 Once again, the candidate has the right ideas, but does not give enough biological detail to get full marks. 2/3

Candidate B

(a) B cells only become active when they meet the specific antigen ✓ to which they are able to respond. If the bacteria are inside a macrophage, then the B cell's receptors won't meet the antigen ✓ on the bacteria. This means that the B cells will not divide to produce plasma cells, ✓ and will not secrete antibodies ✓ against the bacteria.

☑ A good answer. 3/3

(b) The cells only started to take up any bacteria when the particle:macrophage ratio was 1. ✓ On the other hand, they took up glass beads even when the ratio was above 0.01. ✓ When the ratio of particles to macrophages was 10, only about 30% ✓ of the macrophages had taken up bacteria, whereas over 75% of them had taken up glass beads. ✓ This shows the macrophages do take up the bacteria, but not as well as they take up glass beads. ✓

☑ A good answer, which does attempt to 'discuss' by providing statements relating to the relatively low ability of the macrophages to take up the bacteria, but also stating that they do take them up. In general, it is always a good idea to quote data where they are relevant in your answer. 3/3

(c) Normally, lysosomes fuse with phagosomes and release hydrolytic enzymes ✓ into them. These enzymes then hydrolyse (digest) whatever is in the phagosome. ✓ If this doesn't happen, then the bacteria can live inside the phagosome ✓ without being digested.

☑ All correct. 3/3

Question 3

(a) The diagrams show a cell in various stages of the mitotic cell cycle.

A

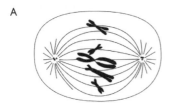

B

C

D

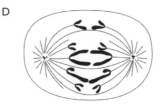

Name the stage represented by each diagram, and arrange them in the correct sequence. (3 marks)

(b) Describe the role of spindle microtubules in mitosis. (3 marks)

(c) The graph below shows the changes in the mass of DNA per cell and total cell mass during two cell cycles. Different vertical scales are used for the two lines.

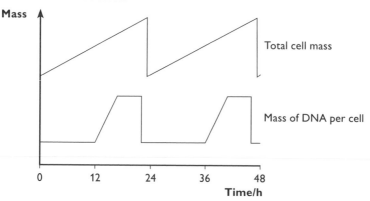

(i) On the graph, write the letter **D** to indicate a time at which **DNA** replication is taking place. (1 mark)

(ii) On the graph, write the letter **C** to indicate a time at which cytokinesis is taking place. (1 mark)

(d) Describe the roles of mitosis in living organisms. (3 marks)

Total: 11 marks

Candidate A

(a) A metaphase, ✓ B prophase, ✓ C telophase, ✓ D anaphase ✓

🖉 The candidate has named each stage correctly, but has not arranged them in the correct order. 2/3

(b) The spindle microtubules pull the chromatids to opposite ends of the cell. ✓

🖉 This is correct, but there is not enough here for three marks. 1/3

(c)

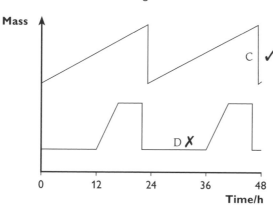

🖉 Cytokinesis is identified correctly, but DNA replication is not. The candidate has written D before the DNA has replicated. 1/2

(d) Mitosis is used in growth and repair. ✓

🖉 This is correct, but not a good enough answer for more than one mark at AS. 1/3

Candidate B

(a) B prophase ✓, A metaphase ✓, D anaphase ✓ C telophase ✓✓

🖉 All identified correctly, and in the right order. 3/3

(b) Spindle microtubules are made by the centrioles. They latch on to the centro-meres ✓of the chromosomes and help them line up on the equator. ✓ Then they pull ✓ on the centromeres so they come apart and they pull the chromatids ✓ to opposite ends of the cell in anaphase.

🖉 A good answer. 3/3

(c)

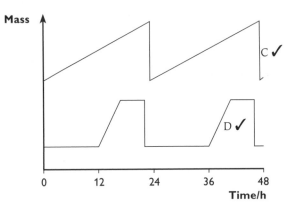

🖉 Both correct. 2/2

(d) Mitosis produces two daughter cells that are genetically identical ✓ to the parent cell. Mitosis is used for growth, or for repairing cells. ✗ It is also used in asexual reproduction. ✓

🖉 The point about producing genetically identical cells is a good one, and it is also correct that mitosis is involved in asexual reproduction. However, the candi-date's second sentence contains an important error. Mitosis cannot repair *cells*. Mitosis can produce new cells, which can help to repair tissues. 2/3

Question 4

The diagrams below show five molecules found in living organisms.

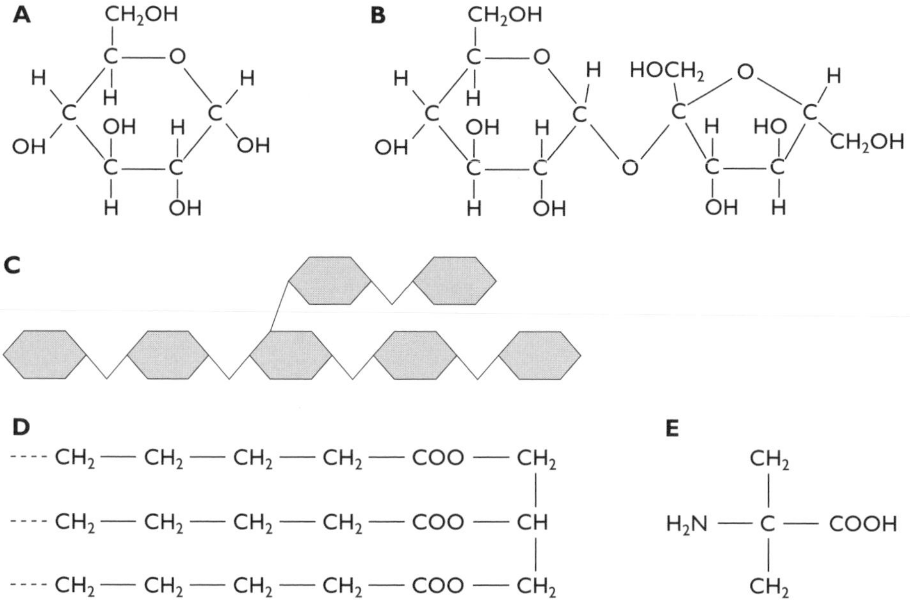

A

B

C

D
```
---- CH₂ — CH₂ — CH₂ — CH₂ — COO — CH₂
                                      |
---- CH₂ — CH₂ — CH₂ — CH₂ — COO — CH
                                      |
---- CH₂ — CH₂ — CH₂ — CH₂ — COO — CH₂
```

E
```
                    CH₂
                     |
        H₂N — C — COOH
                     |
                    CH₂
```

(a)   Give the letter of one molecule that fits each of these descriptions.
        You can use each letter once, more than once or not at all.

  (i)   the form in which carbohydrates are transported through phloem
        tissue in plants   (1 mark)

  (ii) the form in which carbohydrates are stored in animals   (1 mark)

  (iii) a molecule that is insoluble in water   (1 mark)

  (iv) a molecule that links together with others to form a polypeptide (1 mark)

  (v)   a molecule that contains ester bonds   (1 mark)

(b) Explain how the structure of water molecules makes water a good
    solvent.   (3 marks)

                                                       (Total 8 marks)

Candidate A
(a) (i)   A ✗

   🖉 A is a glucose molecule, but plants transport sucrose. Even if you did not know
       what a sucrose molecule looks like, you should know that it is a disaccharide.

   (ii) C ✓

   🖉 Correct. 1/1

(iii) E ✗

🖉 Amino acids are soluble. Either C or D would be correct.

(iv) E ✓

🖉 Correct. 1/1

(v) D ✓

🖉 Correct. 1/1

(b) Water has dipoles and hydrogen bonds, ✓ which help it to dissolve other substances.

🖉 There are no wrong statements in this answer, but it does not really give an explanation of why water is a good solvent — it just states two facts about water molecules. 1/3

Candidate B

(a) (i) B ✓

🖉 Correct. 1/1

(ii) C ✓

🖉 Correct. 1/1

(iii) D or C ✓

🖉 Correct. However, the candidate took an unnecessary risk with (iii), by giving two answers. If the second one had been wrong, it could have negated the first correct one. If you are asked for one answer, it is best to give only one. 1/1

(iv) E ✓

🖉 Correct. 1/1

(v) D ✓

🖉 Correct. 1/1

(b) In a water molecule, the hydrogen atoms have a tiny positive electrical charge and the oxygen atom has a similar negative charge. ✓ Other atoms or ions with electrical charges ✓ are attracted ✓ to these charges on the water molecules. This makes them spread about ✓ among the water molecules.

🖉 This is a good answer. It really does explain how a substance dissolves in water and relates this clearly to the structure of a water molecule. The candidate has actually earned four possible marking points, but there is a maximum of three marks available in total. 3/3

Question 5

The diagram below shows a small part of a human lung as it appears through a microscope.

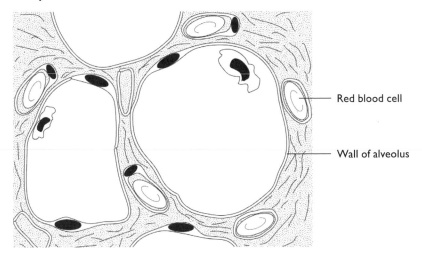

Red blood cell

Wall of alveolus

(a) Name the type of blood vessel in which the red blood cell is present. (1 mark)

(b) Describe and explain two ways in which the structure of the alveoli, shown in the diagram, enables gas exchange to take place rapidly. (4 marks)

(c) Explain why large organisms such as mammals need specialised gas exchange surfaces, whereas small organisms such as a single-celled *Amoeba* do not. (2 marks)

(Total 7 marks)

Candidate A
(a) capillary ✓

    🮑 This is correct. 1/1

(b) They have a large surface area ✓

    They are thin, so oxygen can diffuse across quickly ✓.

    🮑 The statement about a large surface area is correct, but the answer also needs to say *why* this enables gas exchange to take place rapidly (because the question asks you to 'explain'). The second answer is not sufficiently clear — what is thin? It is not the whole alveoli that are thin, but their walls. The second part of this answer does give a clear explanation of why this helps gas exchange to take place quickly. 2/4

(c) Large organisms have small surface areas compared to their volume, ✓ so they need extra surface ✓ to be able to get enough oxygen.

> 🖉 There is a correct and clear statement about surface area to volume ratio, and the answer also just gets a second mark. However, this isn't really very clear — see candidate B for a better explanation. 2/2

Candidate B

(a) capillary ✓

> 🖉 Correct. 1/1

(b) large surface area ✓ — so more oxygen and carbon dioxide molecules can diffuse across at the same time ✓

good supply of oxygen — to maintain a diffusion gradient between the alveoli and the blood

> 🖉 The first way is correct and well explained. However, the second, although true, does not answer the question which is about the *structure of the alveoli*. So just 2/4.

(c) They have small surface area to volume ratios, ✓ but an *Amoeba* has a large surface area to volume ratio. The oxygen that diffuses in across the surface has to supply the whole volume ✓ of the animal, so in a large animal that is not enough and they have specialised gas exchange surfaces to increase the surface area ✓ and let more oxygen diffuse in.

> 🖉 This is a good answer. All the important points are there and it is clearly expressed. 2/2

Question 6

The diagram below shows pressure changes in the left atrium and left ventricle of the heart and the aorta during the cardiac cycle.

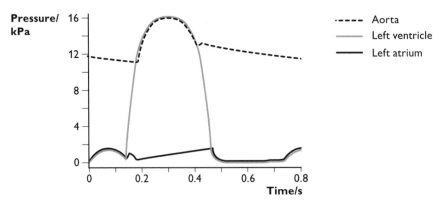

(a) Calculate how many heart beats there will be in one minute. (2 marks)

(b) (i) On the diagram, indicate the point at which the semilunar valves in the aorta snap shut. (1 marks)

(ii) Explain what causes the semilunar valves to shut at this point in the cardiac cycle. (2 marks)

(iii) On the diagram, indicate the period when the left ventricle is contracting. (1 mark)

(iv) On the diagram, draw a line to show the changes in pressure in the right ventricle. (2 marks)

(c) After the blood leaves the heart, it passes into the arteries. The blood pressure gradually reduces and becomes more steady as the blood passes through the arteries.

Explain what causes this reduction and steadying of the blood pressure. (2 marks)

(Total 10 marks)

Candidate A

(a) 1 cycle in 0.75 seconds ✓ so in 60 seconds there will be 60 × 0.75 ✗ = 45 beats

🖉 The student has read the length of one cycle correctly, but the calculation is wrong. 1/2

(b) (i)

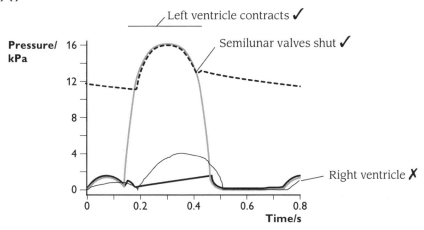

🖉 Correct. 1/1

(ii) The valves shut when the ventricle starts to relax. ✓

🖉 Correct as far as it goes, but it needs to give more information in order to get the second mark. 1/2

(iii) See diagram.

✎ Correct. 1/1

(iv) See diagram

✎ Partly correct. The pressure in the right ventricle is correctly shown as less than that in the left ventricle, but it should be contracting and relaxing at exactly the same times as the left ventricle. 1/2

(c) The pressure gets less as the blood gets further away from the heart. ✓ The muscle in the walls of the arteries contracts ✗ and relaxes to push the blood along, and it does this in between heart beats so the pulse gets evened out.

✎ The first statement is correct, but does not really tell us any more than is in the question. However, it is not correct that the muscles in the artery wall contract and relax to push the blood along. 1/2

Candidate B
(a) 60/0.75 ✓ = 80 beats per minute ✓

✎ Correct. 2/2

(b) (i)

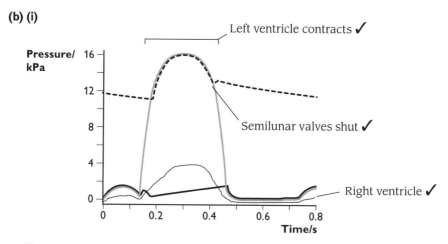

✎ Correct. 1/1

(ii) They close when the pressure of the blood inside the arteries is higher than inside the ventricles ✓ — the blood therefore pushes down on the valves and makes them shut. ✓

✎ Correct. 2/2

(iii) See diagram.

✎ Correct. 1/1

(iv) See diagram.

✎ Correct 2/2

(c) As the blood is forced into the artery as the ventricle contracts, ✓ it pushes outwards on the artery wall, making the elastic tissue stretch. ✓ In between heart beats, the pressure of the blood inside the artery falls, and the elastic tissue recoils. ✓ So the wall keeps expanding and springing back. When it springs back it pushes on the blood in between ✓ heart beats, so this levels up the pressure changes.

🖉 This answer explains very well why the blood pressure levels out. However, it does not mention the overall **fall** in blood pressure. All the same, this is a good answer which gets full marks. 2/2

Question 7

Beetroot cells contain a red pigment that cannot normally escape from the cells through the cell surface membrane.

A student carried out an investigation into the effect of temperature on the permeability of the cell surface membrane of beetroot cells. She measured permeability by using a colorimeter to measure the absorbance of green light by the solutions in which samples of beetroot had been immersed. The greater the absorbance, the more red pigment had leaked out of the beetroot cells

The graph below shows her results.

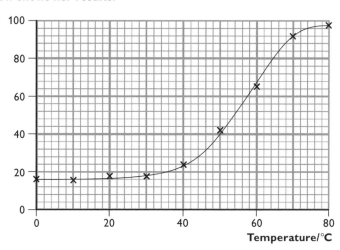

(a) **With reference to the graph, describe the effect of temperature on the absorbance of light in the colorimeter.** (3 marks)

(b) **With reference to the structure of cell membranes, explain the effects you have described in (a).** (4 marks)

(Total 7 marks)

Candidate A

(a) Between 0 and 30 the absorbance goes up very slightly. ✓ Above 40°C it goes up very quickly. ✗ Then it starts to level out at about 70°C. ✓

> The student has correctly identified the three main regions of the graph, stating where changes in gradient occur. However, he or she has used the term 'quickly', which is not correct because the graph does not show anything about time. He or she could also have gained a third mark by quoting some figures from the graph. 2/3

(b) High temperatures damage the proteins in the membrane. ✓ They become denatured, so they leave holes ✓ in the membrane that the beetroot pigment can get through.

> This is a good answer as far as it goes, clearly expressed. However, it is not enough to score a full set of marks. 2/4

Candidate B

(a) The general trend is that the higher the temperature, the greater the absorbance. Between 0 and 30°C, the absorbance increases very slightly ✓ from about 16 to 18 arbitrary units. Above 40°C it increases much more steeply, ✓ levelling off at about 70°C. ✓ The maximum absorbance is 98 arbitrary units.

> A good answer. However, although the candidate has quoted some figures from the graph, he or she has not manipulated them in any way — for example, he or she could have calculated the increase in absorbance between 0 and 30°C. 3/3

(b) As temperature increases, the phospholipids and protein molecules in the membrane move about faster ✓ and with more energy. This leaves gaps in the membrane, so the beetroot pigment molecules can get through ✓ and escape from the cell. The protein molecules start to lose their shape at high temperatures ✓ because their hydrogen bonds break, ✓ so the protein pores get wider ✓ which increases permeability.

> An excellent answer. 4/4

A2
Content
Guidance

L Energy and respiration
Energy in living organisms

ATP

ATP stands for **adenosine triphosphate**. ATP is a phosphorylated nucleotide — it has a similar structure to the nucleotides that make up RNA. However, it has three phosphate groups attached to it instead of one.

An ATP molecule

ATP is used as the energy currency in every living cell. When an ATP molecule is hydrolysed, losing one of its phosphate groups, some of this energy is released and can be used by the cell. In this process, the ATP is converted to ADP (adenosine diphosphate).

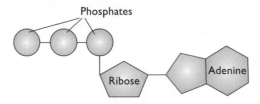

Hydrolysis and formation of ATP

Cells use energy for many different purposes. These include:
- the synthesis of proteins and other large molecules from smaller ones. These are examples of anabolic reactions, that is energy-consuming reactions;
- for active transport of ions and molecules across cell membranes against their concentration gradient (pages 49–50);
- for the transmission of nerve impulses (pages 166–168);
- for movement, for example muscle contraction (such as heart beat, breathing movements, walking) or movement of cilia;
- in mammals and birds, the production of heat to maintain body temperature at a steady level.

Each cell makes its own ATP. The hydrolysis of one ATP molecule releases a small 'packet' of energy that is often just the right size to fuel a particular step in a process. A glucose molecule, on the other hand, would contain far too much energy, so a lot would be wasted if cells used glucose molecules as their immediate source of energy.

All cells make ATP by respiration. This is described in the next few pages.

Respiration

All cells obtain useable energy through respiration. Respiration is the oxidation of energy-containing organic molecules, such as glucose. These are known as **respiratory substrates**. The energy released from this process is used to combine ADP with inorganic phosphate to make ATP.

Respiration may be **aerobic** or **anaerobic**. In both cases, glucose or another respiratory substrate is oxidised.

In aerobic respiration, oxygen is involved, and the substrate is oxidised completely, releasing much of the energy that it contains.

In anaerobic respiration, oxygen is not involved, and the substrate is only partially oxidised. Only a small proportion of the energy it contains is released.

Coenzymes

Respiration involves **coenzymes** called NAD and FAD. A coenzyme is a molecule required for an enzyme to be able to catalyse a reaction. These coenzymes are reduced during respiration. The term 'reduce' means to add hydrogen, so reduced NAD has had hydrogen added to it.

Aerobic respiration

Glucose, $C_6H_{12}O_6$, (or another respiratory substrate) is split to release carbon dioxide as a waste product. The hydrogen from the glucose is combined with atmospheric oxygen. This releases a large amount of energy, which is used to drive the synthesis of ATP.

Glycolysis

Glycolysis is the first stage of respiration. It takes place in the cytoplasm.
- A glucose (or other hexose sugar) molecule is phosphorylated, as two ATPs donate phosphate to it.
- This produces a **hexose bisphosphate** molecule, which splits into two **triose phosphate** molecules.
- Each triose phosphate is converted to a **pyruvate** molecule. This involves the removal of hydrogens, which are taken up by the **coenzyme NAD**. The removal of hydrogens is an **oxidation reaction**. It can also be referred to as **dehydrogenation**. This produces **reduced NAD**. During this step, the phosphate groups from the triose phosphates are added to ADP to produce a small yield of ATP.
- Overall, two molecules of ATP are used and four are made during glycolysis of one glucose molecule, making a net gain of two ATPs per glucose.

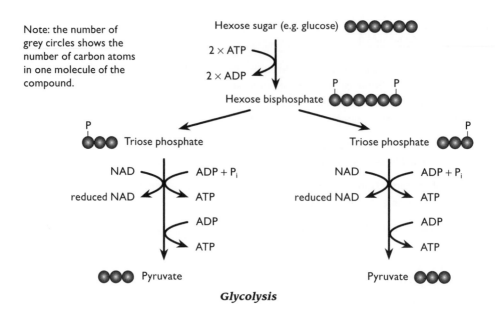

Note: the number of grey circles shows the number of carbon atoms in one molecule of the compound.

Hexose sugar (e.g. glucose)

$2 \times$ ATP

$2 \times$ ADP

Hexose bisphosphate

Triose phosphate

Triose phosphate

NAD → ADP + P_i

reduced NAD → ATP

ADP

ATP

Pyruvate

NAD → ADP + P_i

reduced NAD → ATP

ADP

ATP

Pyruvate

Glycolysis

The link reaction

If oxygen is available, each pyruvate now moves into a mitochondrion, where the **link reaction** and the **Krebs cycle** take place. During these processes, the glucose is completely oxidised.

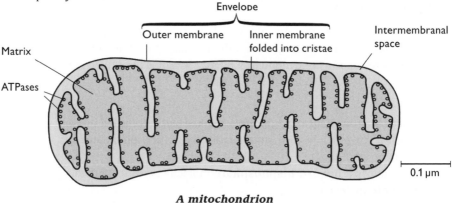

Envelope

Outer membrane   Inner membrane folded into cristae

Intermembranal space

Matrix

ATPases

0.1 µm

A mitochondrion

Carbon dioxide is removed from the pyruvate. This is called **decarboxylation**. This carbon dioxide diffuses out of the mitochondrion and out of the cell.

Hydrogen is also removed from the pyruvate, and is picked up by NAD, producing reduced NAD. This converts pyruvate into a two-carbon compound. This immediately combines with coenzyme A to produce **acetyl coenzyme A**.

The Krebs cycle

Acetyl coenzyme A has two carbon atoms. It combines with a **four-carbon** compound called **oxaloacetate** to produce a **six-carbon** compound, **citrate**. The citrate is gradually converted to the four-carbon compound again through a series of enzyme-controlled steps. These steps all take place in the matrix of the mitochondrion, and each is controlled by specific enzymes.

During this process:
- more carbon dioxide is released (decarboxylation) and diffuses out of the mitochondrion and out of the cell
- more hydrogens are released (dehydrogenation) and picked up by NAD and another coenzyme called FAD. This produces reduced NAD and reduced FAD.

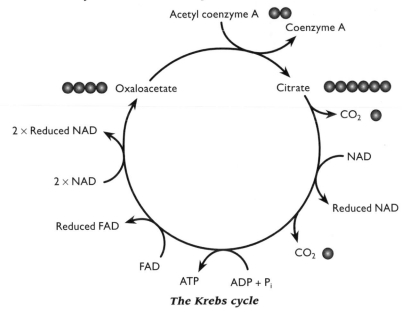

The Krebs cycle

Oxidative phosphorylation

The hydrogens picked up by NAD and FAD are now split into electrons and protons. The electrons are passed along the **electron transport chain**, on the inner membrane of the mitochondrion.

As the electrons move along the chain, they lose energy. This energy is used to actively transport hydrogen ions from the matrix of the mitochondrion, across the inner membrane and into the space between the inner and outer membranes. This builds up a high concentration of hydrogen ions in this space.

The hydrogen ions are allowed to diffuse back into the matrix through special channel proteins that work as ATPases. The movement of the hydrogen ions through the ATPases provides enough energy to cause ADP and inorganic phosphate to combine to make ATP.

At the end of the chain, the electrons reunite with the protons from which they were originally split. They combine with oxygen to produce water. This is why oxygen is required in aerobic respiration — it acts as the final acceptor for the hydrogens removed from the respiratory substrate during glycolysis, the link reaction and the Krebs cycle.

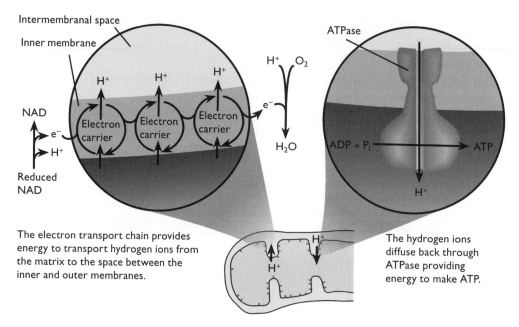

Oxidative phosphorylation

The electron transport chain provides energy to transport hydrogen ions from the matrix to the space between the inner and outer membranes.

The hydrogen ions diffuse back through ATPase providing energy to make ATP.

Anaerobic respiration

If oxygen is not available, oxidative phosphorylation cannot take place, as there is nothing to accept the electrons and protons at the end of the electron transport chain. This means that reduced NAD is not reoxidised, so the mitochondrion quickly runs out of NAD or FAD that can accept hydrogens from the Krebs cycle reactions. The Krebs cycle and the link reaction therefore come to a halt.

Glycolysis, however, can still continue, so long as the pyruvate produced at the end of it can be removed and the reduced NAD can be converted back to NAD.

The lactate pathway

In mammals, the pyruvate is removed by converting it to **lactate**.

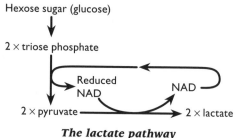

The lactate pathway

The lactate that is produced (usually in muscles) diffuses into the blood and is carried in solution in the blood plasma to the liver. Here, liver cells convert it back to pyruvate. This requires oxygen, so extra oxygen is required after exercise has finished. The extra oxygen is known as the **oxygen debt**. Later, when the exercise has finished and oxygen is available again, some of the pyruvate in the liver cells is oxidised through the link reaction, the Krebs cycle and the electron transport chain. Some of the pyruvate is reconverted to glucose in the liver cells. The glucose may be released into the blood or converted to glycogen and stored.

The ethanol pathway

In yeast and in plants, the pyruvate is removed by converting it to ethanol.

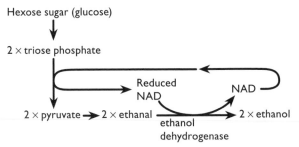

The ethanol pathway

ATP yield in aerobic and anaerobic respiration

Only small amounts of ATP are produced when one glucose molecule undergoes anaerobic respiration. This is because only glycolysis is completed. The Krebs cycle and oxidative phosphorylation, which produce most ATP, do not take place.

The precise number of molecules of ATP produced in aerobic respiration of one glucose molecule varies between different organisms and different cells, but is usually between 30 and 32 molecules. The diagram on page 144 shows ATP yields in respiration.

Respiratory substrates

Glucose is not the only respiratory substrate. All carbohydrates, lipids and proteins can also be used as respiratory substrates.

Energy values of different respiratory substrates

Respiratory substrate	Energy released/kJ g^{-1}
carbohydrate	16
lipid	39
protein	17

You can see that lipid provides more than twice as much energy per gram as carbohydrate or protein. This is because a lipid molecule contains relatively more hydrogen atoms (in comparison with carbon or oxygen atoms) than carbohydrate or protein molecules do. You have seen that it is hydrogen atoms that are used to generate ATP via the electron transport chain.

Many cells in the human body are able to use a range of different respiratory substrates. However, brain cells can only use glucose. Heart muscle preferentially uses fatty acids.

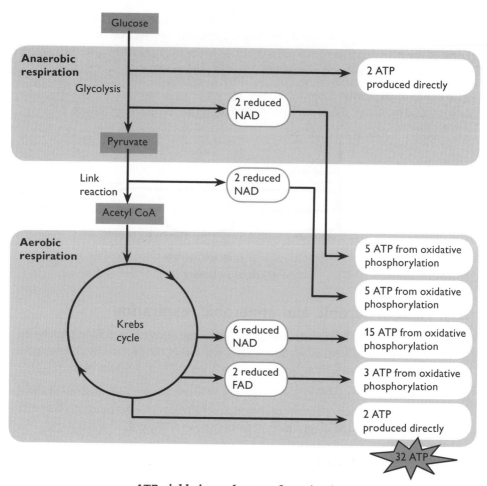

ATP yields in each step of respiration

Respiratory quotients

It is possible to get a good idea of which respiratory substrate the cells in an organism are using by measuring the volume of oxygen it is taking in and the volume of carbon dioxide it is giving out.

The **respiratory quotient, RQ** is $\dfrac{\text{volume of } CO_2 \text{ given out}}{\text{volume of } O_2 \text{ taken in}}$

RQs for different substrates undergoing aerobic respiration

Respiratory substrate	RQ
Carbohydrate	1.0
Lipid	0.7
Protein	0.9

The values in the table are for aerobic respiration. If a cell or an organism is respiring anaerobically, then no oxygen is being used. The RQ is therefore infinity (∞).

Using respirometers

There are various different types of respirometer. One type is shown in the diagram.

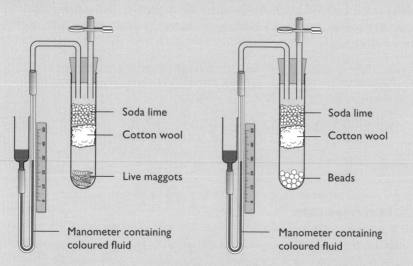

- Soda lime
- Cotton wool
- Live maggots
- Manometer containing coloured fluid

- Soda lime
- Cotton wool
- Beads
- Manometer containing coloured fluid

A respirometer

Using a respirometer to measure the rate of uptake of oxygen

The organisms to be investigated are placed in one tube, and non-living material of the same mass in the other tube. Soda lime is placed in each tube, to absorb all carbon dioxide. Cotton wool prevents contact of the soda lime with the organisms.

Coloured fluid is poured into the reservoir of each manometer and allowed to flow into the capillary tube. It is essential that there are no air bubbles. You must end up with exactly the same quantity of fluid in the two manometers.

Two rubber bungs are now taken, fitted with tubes as shown in the diagram. Close the spring clips. Attach the manometers to the bent glass tubing, ensuring an airtight connection. Next, place the bungs into the tops of the tubes.

Open the spring clips. (This allows the pressure throughout the apparatus to equilibrate with atmospheric pressure.) Note the level of the manometer fluid in each tube. Close the clips. Each minute, record the level of the fluid in each tube.

As the organisms respire, they take oxygen from the air around them and give out carbon dioxide. The removal of oxygen from the air inside the tube reduces the volume and pressure, causing the manometer fluid to move towards the organisms. The carbon dioxide given out is absorbed by the soda lime. The distance moved by the fluid is therefore affected only by the oxygen taken up and not by the carbon dioxide given out.

You would not expect the manometer fluid in the tube with no organisms to move, but it may do so because of temperature changes. This allows you to control for this variable, by subtracting the distance moved by the fluid in the control manometer from the distance moved in the experimental manometer (connected to the living organisms), to give you an adjusted distance moved.

Calculate the mean (adjusted) distance moved by the manometer fluid per minute. If you know the diameter of the capillary tube, you can convert the distance moved to a volume:

volume of liquid in a tube = length × πr^2

This gives you a value for the volume of oxygen absorbed by the organisms per minute.

Using a respirometer to investigate the effect of temperature on the rate of respiration

The respirometer can be placed in water baths at different temperatures. You can use the same respirometer for the whole experiment. Or you could have different ones for each temperature. (In each case, there are difficulties with controlling some variables — you might like to think about what these are.) At each temperature, you need a control respirometer with no organisms in it.

If you are simply *comparing* the rates of respiration at different temperatures, then you do not need to convert the distance moved by the manometer fluid to a volume. You could just plot distance moved on the *y*-axis of your graph and time on the *x*-axis.

The rate of respiration is represented by the gradient of the graph.

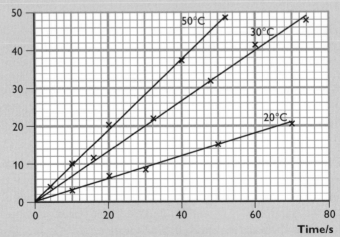

Comparing rates of respiration at different temperatures

At 50 °C manometer fluid travelled 47 mm in 50 s

Rate of respiration = $\frac{47}{50}$ = 0.94 mm s^{-1}

At 30 °C manometer fluid travelled 40 mm in 60 s

Rate of respiration = $\frac{40}{60}$ = 0.67 mm s^{-1}

At 20 °C manometer fluid travelled 21 mm in 70 s

Rate of respiration = $\frac{24}{70}$ = 0.34 mm s^{-1}

Using a respirometer to measure RQ

For this, we need to know both how much oxygen is taken in, and how much carbon dioxide is given out.

Set up two respirometers as shown in the diagram on page 145. However, the second respirometer should also contain the same mass of live maggots (or whatever organism you are investigating) but should **not** contain soda lime. You could put some inert material into the tube (for example, the beads) instead of soda lime. The mass and volume of the inert material should be the same as the mass and volume of the soda lime.

This second tube is therefore just like the first one except that it does not contain soda lime. The carbon dioxide given out by the respiring organisms is therefore not absorbed.

The difference between the distance moved by the manometer fluid in the experimental tube and the distance moved in the control tube is therefore due to the carbon dioxide given out.

distance moved by fluid in experimental tube = x mm

distance moved by fluid in control tube = y mm

x mm represents the oxygen taken up

$x - y$ represents the carbon dioxide given out

therefore RQ = $\frac{x - y}{x}$

For example, if the respiratory substrate is carbohydrate, then the amount of carbon dioxide given out will equal the amount of oxygen taken in. The fluid in the control tube will not move, so $y = 0$.

RQ is then $\frac{x}{x} = 1$

M Photosynthesis

An overview of photosynthesis

Photosynthesis is a series of reactions in which energy transferred as light is transformed to chemical energy. Energy from light is trapped by chlorophyll, and this energy is then used to

- split apart the strong bonds in water molecules to release hydrogen
- produce ATP
- reduce a substance called NADP.

NADP stands for nicotinamide adenine dinucleotide phosphate, which — like NAD — is a coenzyme.

The ATP and reduced NADP are then used to add hydrogen to carbon dioxide, to produce carbohydrate molecules such as glucose. These complex organic molecules contain some of the energy that was originally in the light. The oxygen from the split water molecules is a waste product, and is released into the air.

There are many different steps in photosynthesis, which can be divided into two main stages — the light-dependent reactions and the light-independent reactions.

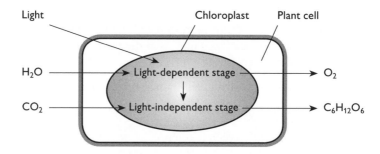

An overview of photosynthesis

Chloroplasts

Photosynthesis takes place inside **chloroplasts**. These are organelles surrounded by two membranes, called an **envelope**. They are found in mesophyll cells in leaves. Palisade mesophyll cells contain most chloroplasts but they are also found in spongy mesophyll cells. Guard cells also contain chloroplasts. You can see a diagram of the structure of a leaf on page 152.

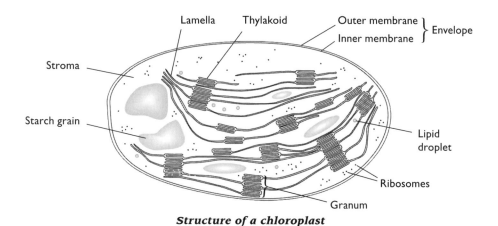

Structure of a chloroplast

The membranes inside a chloroplast are called **lamellae**, and it is here that the light-dependent reactions take place. The membranes contain **chlorophyll** molecules, arranged in groups called **photosystems**. There are two kinds of photosystems, PSI and PSII, each of which contains slightly different kinds of chlorophyll.

There are enclosed spaces between pairs of membranes, forming fluid-filled sacs called **thylakoids**. These are involved in photophosphorylation — the formation of ATP using energy from light. Thylakoids are often arranged in stacks called **grana** (singular: granum).

The 'background material' of the chloroplast is called the **stroma**, and this is where the light-independent reactions take place.

Chloroplasts often contain starch grains and lipid droplets. These are stores of energy-containing substances that have been made in the chloroplast but are not immediately needed by the cell or by other parts of the plant.

The light-dependent reactions

Chlorophyll molecules in PSI and PSII absorb light energy. The energy excites electrons, raising their energy level so that they leave the chlorophyll. The chlorophyll is said to be photo-activated.

PSII contains an enzyme that splits water when activated by light. This reaction is called **photolysis** ('splitting by light'). The water molecules are split into oxygen and hydrogen atoms. Each hydrogen atom then loses its electron, to become a positively charged hydrogen ion (proton), H^+. The electrons are picked up by the chlorophyll in PSII, to replace the electrons they lost. The oxygen atoms join together to form oxygen molecules, which diffuse out of the chloroplast and into the air around the leaf.

$$2H_2O \xrightarrow{\text{light}} 4H^+ + 4e^- + O_2$$

The electrons emitted from PSII are picked up by electron carriers in the membranes of the thylakoids. They are passed along a chain of these carriers, losing energy as they go. The energy they lose is used to make ADP combine with a phosphate group, producing ATP. This is called **photophosphorylation**. At the end of the electron carrier chain, the electron is picked up by PSI, to replace the electron the chlorophyll in PSI had lost.

The electrons from PSI are passed along a different chain of carriers to NADP. The NADP also picks up the hydrogen ions from the split water molecules. The NADP becomes reduced NADP.

We can show all of this in a diagram called the **Z-scheme**. The higher up the diagram, the higher the energy level. If you follow one electron from a water molecule, you can see how it

- is taken up by PSII
- has its energy raised as the chlorophyll in PSII absorbs light energy
- loses some of this energy as it passes along the electron carrier chain
- is taken up by PSI
- has its energy raised again as the chlorophyll in PSI absorbs light energy
- becomes part of a reduced NADP molecule

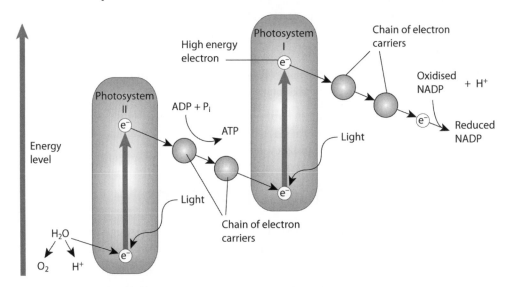

Summary of the light-dependent reactions of photosynthesis — the Z-scheme

At the end of this process, two new substances have been made. These are ATP and reduced NADP. Both of them will now be used in the next stage of photosynthesis, the light-independent reactions.

Non-cyclic and cyclic photophosphorylation

The sequence of events just described and shown in the flow diagram above is known as **non-cyclic photophosphorylation**.

There is an alternative pathway for the electron that is emitted from PSI. It can simply be passed along the electron transport chain, then back to PSI again. ATP is produced as it moves along the electron transport chain (photophosphorylation). However, no reduced NADP is produced. This is called **cyclic photophosphorylation**.

The light-independent reactions

These take place in the stroma of the chloroplast, where the enzyme ribulose bisphosphate carboxylase, usually known as **rubisco**, is found.

Carbon dioxide diffuses into the stroma from the air spaces within the leaf. It enters the active site of rubisco, which combines it with a 5-carbon compound called ribulose bisphosphate, **RuBP**. The products of this reaction are two 3-carbon molecules, glycerate 3-phosphate, **GP**. The combination of carbon dioxide with RuBP is called **carbon fixation**.

Energy from ATP and hydrogen from reduced NADP are then used to convert the GP into **triose phosphate**, **TP**. Triose phosphate is the first carbohydrate produced in photosynthesis.

Most of the triose phosphate is used to produce ribulose bisphosphate, so that more carbon dioxide can be fixed. The rest is used to make glucose or whatever other organic substances the plant cell requires. These include polysaccharides such as starch for energy storage and cellulose for making cell walls, sucrose for transport, amino acids for making proteins, lipids for energy storage and nucleotides for making DNA and RNA.

This cyclical series of reactions is known as the **Calvin cycle**.

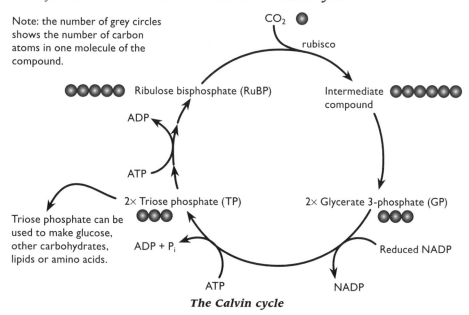

Note: the number of grey circles shows the number of carbon atoms in one molecule of the compound.

The Calvin cycle

Leaf structure and photosynthesis

The diagram shows how various features of a leaf are adaptations that enable photosynthesis to take place.

The overall shape of most leaves is thin and broad. The thinness allows sunlight to pass through to all the palisade and spongy cells. The broad shape maximises the surface area for absorption of sunlight and carbon dioxide.

The epidermal cells have no chloroplasts, so light passes through and reaches the palisade cells. They secrete a waxy cuticle, which reduces the evaporation of water from the upper surface of the leaf.

Most photosynthesis takes place in the palisade layer, as this is where the cells contain most chloroplasts. The position of this layer close to the surface of the leaf enables sunlight to reach it easily. The tall, narrow cells mean light has to pass through only three cell walls to reach the chloroplasts. The cells are tightly packed to capture the maximum sunlight.

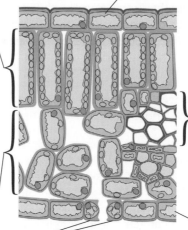

Vascular bundles contain xylem vessels and phloem sieve tubes. Xylem helps to support the leaf, holding it out flat so it can absorb sunlight. Xylem vessels bring water to the leaf, which is necessary to keep cells turgid as well as being a reactant in photosynthesis. Phloem sieve tubes take away assimilates such as sucrose, made by photosynthesis.

Some photosynthesis takes place in the spongy layer, as there are some chloroplasts in these cells. The large air spaces between them allow easy and rapid diffusion of carbon dioxide from the stomata to the chloroplasts in the palisade layer and the spongy layer.

Two guard cells surround each stoma. The stomata allow rapid diffusion of carbon dioxide into the leaf for photosynthesis, and they connect with the air spaces inside the spongy layer. When the guard cells are turgid, they bend outwards and leave a space between them — the stomatal pore. When they are flaccid, they are less bent and the pore between them is closed.

The lower epidermal cells have no chloroplasts. They may secrete a waxy cuticle to cut down water loss, but this is usually thinner than on the upper surface.

Limiting factors in photosynthesis

The rate at which photosynthesis takes place is directly affected by several environmental factors.

- Light intensity. This affects the rate of the light-dependent reaction, because this is driven by energy transferred in light rays.
- Temperature. This affects the rate of the light-independent reaction. At higher temperatures, molecules have more kinetic energy so collide more often and are more likely to react when they do collide. (The rate of the light-dependent

reaction is not affected by temperature, as the energy that drives this process is light energy, not heat energy.)

- Carbon dioxide concentration in the atmosphere. Carbon dioxide is a reactant in photosynthesis. Normal air contains only about 0.04% carbon dioxide.
- Availability of water. Water is a reactant in photosynthesis, but there is usually far more water available than carbon dioxide, so even if water supplies are low this is not usually a problem. However, water supply can affect the rate of photosynthesis *indirectly*, because a plant that is short of water will close its stomata, preventing carbon dioxide from diffusing into the leaf.

If the level of any one of these factors is too low, then the rate of photosynthesis will be reduced. The factor that has the greatest effect in reducing the rate is said to be the **limiting factor**.

Over this range of the graphs, light intensity is the limiting factor. If light intensity increases, then the rate of photosynthesis increases.

Over this range of the graph, light intensity is not a limiting factor. If light intensity increases, there is no effect on the rate of photosynthesis. Some other factor is limiting the rate.

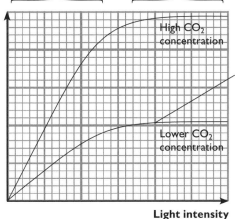

Here, carbon dioxide is the limiting factor. We can tell this is so because when the concentration of carbon dioxide is increased, the rate of photosynthesis increases (see top curve).

Limiting factors for photosynthesis

Investigating the effect of environmental factors on the rate of photosynthesis

One way to measure the rate of photosynthesis is to measure the rate at which oxygen is given off by an aquatic plant. There are various ways in which oxygen can be collected and measured. One method is shown in the diagram overleaf.

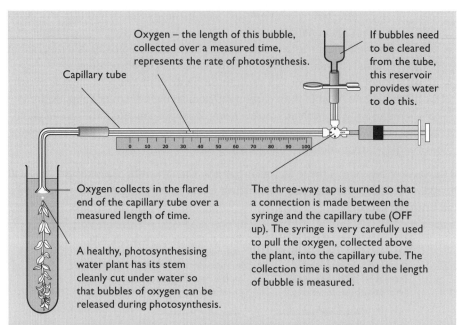

Oxygen – the length of this bubble, collected over a measured time, represents the rate of photosynthesis.

If bubbles need to be cleared from the tube, this reservoir provides water to do this.

Capillary tube

Oxygen collects in the flared end of the capillary tube over a measured length of time.

The three-way tap is turned so that a connection is made between the syringe and the capillary tube (OFF up). The syringe is very carefully used to pull the oxygen, collected above the plant, into the capillary tube. The collection time is noted and the length of bubble is measured.

A healthy, photosynthesising water plant has its stem cleanly cut under water so that bubbles of oxygen can be released during photosynthesis.

Apparatus for measuring the rate of photosynthesis

Alternatively, you can make calcium alginate balls containing green algae and place them in hydrogencarbonate indicator solution. As the algae photosynthesise, they take in carbon dioxide which causes the pH around them to increase. The indicator changes from orange, through red to magenta. You can find details of this technique at **http://www-saps.plantsci.cam.ac.uk/worksheets/ssheets/ssheet23.htm**

Whichever technique is used, you should change one factor (your independent variable) while keeping all others constant (the control variables). The dependent variable will be the rate at which oxygen is given off (measured by the volume of oxygen collected per minute in the capillary tube) *or* the rate at which carbon dioxide is used (measured by the rate of change of colour of the hydrogencarbonate indicator solution).

The independent variables you could investigate are:
- Light intensity. You can vary this by using a lamp to shine light onto the plant or algae. The closer the lamp, the higher the light intensity.
- Wavelength of light. You can vary this by placing coloured filters between the light source and the plant. Each filter will allow only light of certain wavelengths to pass through.

- Carbon dioxide concentration. You can vary this by adding sodium hydrogencarbonate to the water around the aquatic plant. This contains hydrogencarbonate ions, which are used as a source of carbon dioxide by aquatic plants.
- Temperature. The part of the apparatus containing the plant or algae can be placed in a water bath at a range of controlled temperatures.

Chloroplast pigments

A **pigment** is a substance that absorbs light of some wavelengths but not others. The wavelengths that is does **not** absorb are reflected from it.

Chlorophyll is the main pigment contained in chloroplasts. It looks green because it reflects green light. Other wavelengths (colours) of light are absorbed.

The diagram shows the wavelengths of light absorbed by the various pigments found in chloroplasts. These graphs are called **absorption spectra**.

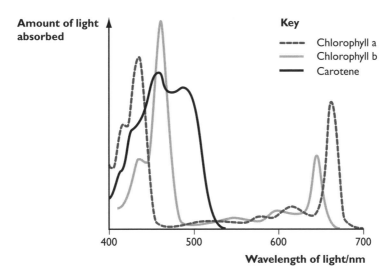

Absorption spectra for chloroplast pigments

If we shine light of various wavelengths on chloroplasts containing different pigments, we can measure the rate at which they give off oxygen. These graphs are called **action spectra**.

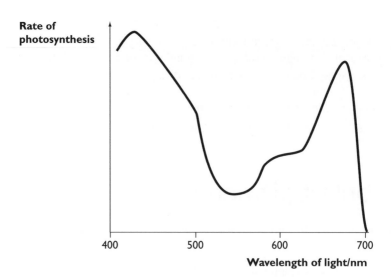

Action spectrum for chloroplast pigments

Chlorophyll a is the most abundant pigment in most plants. Its absorption peaks are 430 nm (blue) and 662 nm (red). It emits an electron when it absorbs light.

A chlorophyll a molecule

Chlorophyll b is similar to chlorophyll a, but its absorption peaks are 453 nm and 642 nm. It has a similar role to chlorophyll a, but is not as abundant.

Carotenoids are **accessory pigments**. They are orange pigments that protect chlorophyll from damage by the formation of single oxygen atoms (free radicals). They can also absorb wavelengths of light that chlorophyll cannot absorb, and pass on some of the energy from the light to chlorophyll.

Xanthophylls are also accessory pigments, capturing energy from wavelengths of light that are not absorbed by chlorophyll.

Separating chlorophyll pigments by chromatography

Chromatography is a method of separation that relies on the different solubilities of different solutes in a solvent. A mixture of chlorophyll pigments is dissolved in a solvent, and then a small spot is placed onto chromatography paper. The solvent gradually moves up the paper, carrying the solutes with it. The more soluble the solvent, the further up the paper it is carried.

There are various methods. The one described here uses thin layer chromatography on specially prepared strips instead of paper. Only an outline of the procedure is given here, so you cannot use these instructions to actually carry out the experiment. You can find more details about this technique at **http://www-saps.plantsci.cam.ac.uk/worksheets/ssheets/ssheet10.htm**. Your teacher should give you an opportunity to do a chromatography experiment, perhaps using a different technique from the one described here.

Cut a TLC plate into narrow strips, about 1.25 cm wide, so they fit into a test tube. Do not put your fingers on the powdery surface.

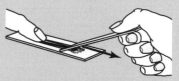

Put 2 or 3 grass leaves on a slide. Using another slide scrape the leaves to extract cell contents.

Add 6 drops of propanone to the extract and mix.

Transfer the mixture to a watch glass. Allow this to dry out almost completely – a warm air flow will speed this up.

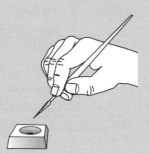

Transfer tiny amounts of the concentrated extract onto a spot 1 cm from one end of the TLC strip.

Touch very briefly with the fine tip of the brush and let that spot dry before adding more. Keep the spot to 1 mm diameter if you can. The final spot, called the origin, should be small but dark green.

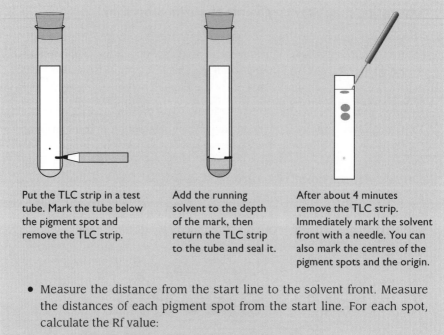

Put the TLC strip in a test tube. Mark the tube below the pigment spot and remove the TLC strip.

Add the running solvent to the depth of the mark, then return the TLC strip to the tube and seal it.

After about 4 minutes remove the TLC strip. Immediately mark the solvent front with a needle. You can also mark the centres of the pigment spots and the origin.

- Measure the distance from the start line to the solvent front. Measure the distances of each pigment spot from the start line. For each spot, calculate the Rf value:

$$Rf = \frac{\text{distance from start line to pigment spot}}{\text{distance from start line to solvent front}}$$

- You can use the Rf values to help you to identify the pigments. Rf values differ depending on the solvent you have used, but typical values might be:

chlorophyll a	0.60
chlorophyll b	0.50
carotene	0.95
xanthophyll	0.35

You may also see a small grey spot with an Rf value of about 0.8. This is phaeophytin, which is not really a chlorophyll pigment, but is a breakdown product generated during the extraction process.

N Regulation and control
Excretion

Excretion is the removal of waste products generated by metabolic reactions inside body cells. Some of these products are toxic, while others are simply in excess of requirements.

In mammals, the two major excretory products are:

- **carbon dioxide**, produced by aerobic respiration. Carbon dioxide dissolves in water to produce a weak acid, so if too much builds up in body fluids the pH drops, which can damage cells and disrupt metabolism. Carbon dioxide is transported to the lungs dissolved in blood plasma and excreted in expired air.
- **nitrogenous excretory products**, in particular **urea**. Excess amino acids cannot be stored in the body. In the liver, they are converted to urea, $CO(NH_2)_2$, and a keto acid. The keto acid can be respired to provide energy, or converted to fat for storage. The urea dissolves in the blood plasma and is removed and excreted by the kidneys.

The structure and histology of kidneys

Each kidney is supplied with oxygenated blood through a renal artery. Blood is removed in the renal vein. A tube called the **ureter** takes urine from the kidney to the bladder.

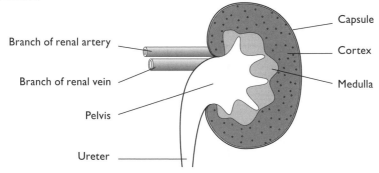

Section through a kidney and its associated vessels

Each kidney contains thousands of microscopic tubes called **nephrons**. The beginning of each nephron is a cup-shaped structure called a **renal capsule**. This is in the **cortex** of the kidney. The tube leads from the renal capsule down into the kidney **medulla**, then loops back into the cortex before finally running back down through the medulla into the **pelvis** of the kidney, where it joins the ureter.

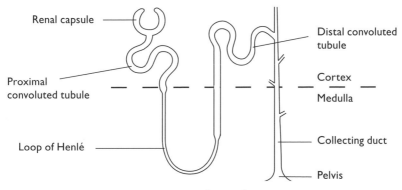

Structure of a nephron

Each nephron has a network of blood vessels associated with it. Blood arrives in the afferent arteriole (from the renal artery), and is delivered to a network of capillaries, called a glomerulus, in the cup of the renal capsule. Blood leaves the glomerulus in the efferent arteriole, which is narrower than the afferent arteriole. This leads to another network of capillaries that wraps around the nephron, before delivering the blood to a branch of the renal vein.

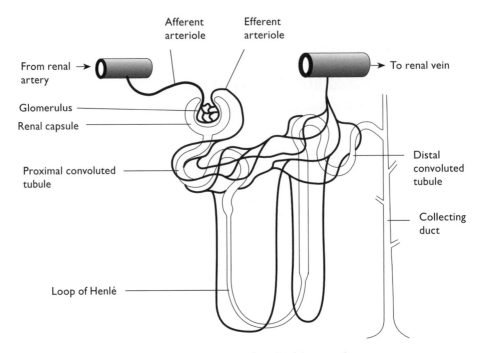

The blood vessels associated with a nephron

The diagrams below show the histology of the kidney. Histology is the structure of tissues.

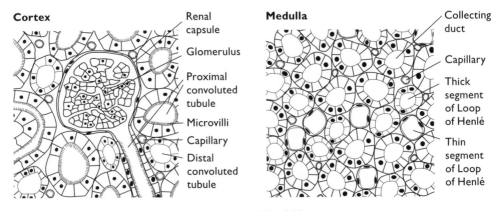

Histology of the kidney

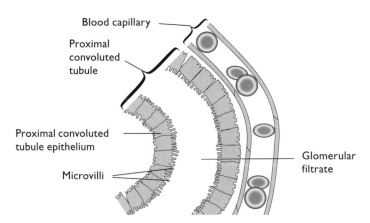

Longitudinal section of part of a proximal convoluted tubule

How urine is produced in a nephron

Ultrafiltration

The blood in a glomerulus is separated from the space inside the renal capsule by:

- the capillary wall (**endothelium**) which is one cell thick and has pores in it;
- the **basement membrane** of the wall of the renal capsule;
- the layer of cells making up the wall of the renal capsule, called **podocytes**; these cells have slits between them.

The blood in a glomerulus is at a relatively high pressure, because the efferent arteriole is narrower than the afferent arteriole. This forces molecules from the blood through these three structures, into the renal capsule. The pores in the capillary endothelium and the slits between the podocytes will let all molecules through, but the basement membrane acts as a filter and will only let small molecules pass through.

Substances that can pass through include water, glucose, inorganic ions such as Na^+, K^+ and Cl^- and urea.

Substances that cannot pass through include red and white blood cells and plasma proteins (such as albumen and fibrinogen).

The liquid that seeps through into the renal capsule is called glomerular filtrate.

Comparison of the composition of blood and glomerular filtrate

Component	Blood	Glomerular filtrate
Cells	Contains red cells, white cells and platelets	No cells
Water/g dm^{-3}	900	900
Inorganic ions (including Na^+, K^+ and Cl^-)/g dm^{-3}	7	7
Plasma proteins/g dm^{-3}	45	0
Glucose/g dm^{-3}	1	1
Urea/g dm^{-3}	0.3	0.3

Selective reabsorption in the proximal convoluted tubule

Some of the substances that are filtered into the renal capsule need to be retained by the body. These include:

- much of the water
- all of the glucose
- some of the inorganic ions

These substances are therefore taken back into the blood through the walls of the proximal convoluted tubule. This is called **selective reabsorption**.

The cells in the walls of the tubule have many mitochondria, to provide ATP for active transport. Their surfaces facing the lumen of the tubule have a large surface area provided by microvilli.

- **Active transport** is used to move Na^+ out of the outer surface of a cell in the wall of the proximal convoluted tubule, into the blood.
- This lowers the concentration of Na^+ inside the cell, so that Na^+ ions diffuse *into* the cell from the fluid inside the tubule. The Na^+ ions diffuse through protein transporters in the cell surface membrane of the cell.
- As the Na^+ ions diffuse through these transporter proteins, they carry glucose molecules with them. This is called **co-transport**. The glucose molecules move through the cell and diffuse into the blood.
- The movement of Na^+ and glucose into the blood decreases the water potential in the blood. Water therefore moves by **osmosis** from the fluid inside the tubule, down a water potential gradient through the cells making up the wall of the tubule and into the blood.

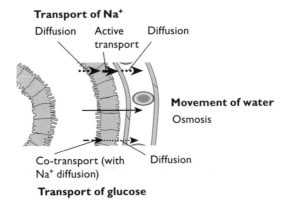

Selective reabsorption in the proximal convoluted tubule

As a result, the fluid inside the nephron now has:

- no glucose
- a lower concentration of Na^+ than the filtrate originally had
- less water than the filtrate originally had

About 50% of the urea is also reabsorbed in the proximal convoluted tubule.

The loop of Henlé

Some, but not all, nephrons have long loops of Henlé that dip down into the medulla and then back up into the cortex. The function of the loop of Henlé is to build up a high concentration of Na^+ and Cl^- in the tissues of the medulla. This allows highly concentrated urine to be produced. Note that the loop of Henlé itself does not produce highly concentrated urine.

As fluid flows down the descending limb of the loop of Henlé, water moves out of it by osmosis. By the time the fluid reaches the bottom of the loop, it has a much lower water potential than at the top of the loop. As it flows up the ascending limb, Na^+ and Cl^- move out of the fluid into the surrounding tissues, first by diffusion and then by active transport.

This creates a low water potential in the tissues of the medulla. The longer the loop, the lower the water potential that can be produced.

The distal convoluted tubule and collecting duct

The fluid inside the tubule as it leaves the loop of Henlé and moves into the collecting duct has lost a little more water and more Na^+ than it had when it entered the loop. Because more water has been lost, the concentration of urea has increased.

Now, in the distal convoluted tubule, Na^+ is actively transported out of the fluid.

The fluid then flows through the collecting duct. This passes through the medulla, where you have seen that a low water potential has been produced by the loop of Henlé. As the fluid continues to flow through the collecting duct, water moves down the water potential gradient from the collecting duct and into the tissues of the medulla. This further increases the concentration of urea in the tubule. The fluid that finally leaves the collecting duct and flows into the ureter is **urine**.

Osmoregulation

Osmoregulation is the control of the water content of body fluids. It is part of **homeostasis**, the maintenance of a constant internal environment. It is important that cells are surrounded by tissue fluid of a similar water potential to their own contents, to avoid too much water loss or gain which could disrupt metabolism.

You have seen that water is lost from the fluid inside a nephron as it flows through the collecting duct. The permeability of the walls of the distal convoluted tubule and collecting duct can be varied.

- If they are permeable, then much water can move out of the tubule and the urine becomes concentrated. The water is taken back into the blood and retained in the body.
- If they are made impermeable, little water can move out of the tubule and the urine remains dilute. A lot of water is removed from the body.

ADH

ADH is **antidiuretic hormone**. It is secreted from the **anterior pituitary gland** into the blood.

When the water potential of the blood is too low (that is, it has too little water in it), this is sensed by **osmoreceptor cells** in the **hypothalamus**. The osmoreceptor cells are neurones (nerve cells). They produce ADH, which moves along their axons and into the anterior pituitary gland from where it is secreted into the blood.

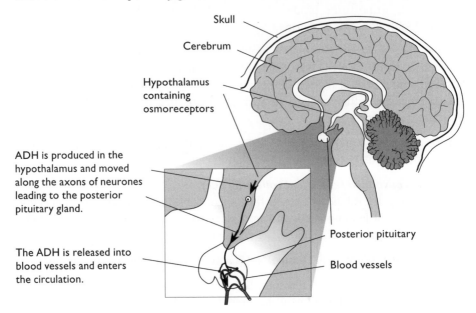

Osmoreceptors and the secretion of ADH

The ADH travels in solution in the blood plasma. When it reaches the walls of the collecting duct, it makes them permeable to water. Water is therefore reabsorbed from the fluid in the collecting duct and small volumes of concentrated urine are produced.

When the water potential of the blood is too high (that is, it has too much water in it), this is sensed by the osmoreceptor cells and less ADH is secreted. The collecting duct walls therefore become less permeable to water and less is reabsorbed into the blood. Large volumes of dilute urine are produced.

Negative feedback
The mechanism for controlling the water content of the body, using ADH, is an example of **negative feedback**.

When the water potential of the blood rises too high or falls too low, this is sensed by **receptor** cells. They cause an action to be taken by **effectors** which cause the water potential to be moved back towards the correct value.

In this case, the receptors are the osmoreceptor cells in the hypothalamus, and the effectors are their endings in the anterior pituitary gland that secrete ADH.

You can read about negative feedback in the control of blood glucose concentration on page 169 to 170.

The mammalian nervous system

Coordination

In multicellular organisms, such as plants and animals, it is essential that cells can communicate with each other. This allows them to coordinate their activities appropriately. Organisms have specialised cells or molecules, called **receptors**, which are sensitive to changes in their internal or external environment. These trigger events in the organism that bring about coordinated responses to the environmental changes.

Neurones

Neurones (nerve cells) are highly specialised cells that are adapted for the rapid transmission of electrical impulses, called **action potentials**, from one part of the body to another.

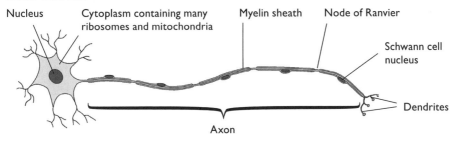

The structure of neurones

Information picked up by a **receptor** is transmitted to the **central nervous system** (brain or spinal cord) as action potentials travelling along a **sensory neurone**. These neurones have their cell bodies in small swellings, called **ganglia**, just outside the **spinal cord**. The impulse may then be transmitted to a **relay neurone**, which lies entirely within the brain or spinal cord. The impulse is then transmitted to many other neurones, one of which may be a **motor neurone**. This has its cell body within the central nervous system, and a long axon which carries the impulse all the way to an **effector** (a muscle or gland).

In some cases, the impulse is sent on to an effector before it reaches the 'conscious' areas of the brain. The response is therefore automatic, and does not involve any decision-making. This type of response is called a **reflex**, and the arrangement of neurones is called a **reflex arc**.

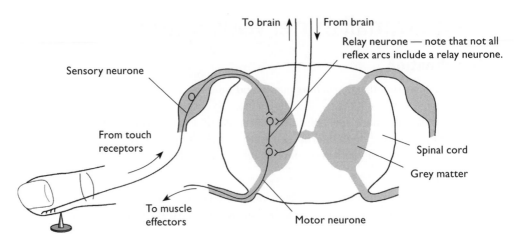

To brain   From brain

Relay neurone — note that not all
reflex arcs include a relay neurone.

Sensory neurone

From touch
receptors

Spinal cord

Grey matter

To muscle
effectors

Motor neurone

Arrangement of neurones in a reflex arc

The **myelin sheath** is made up of many layers of cell membranes of **Schwann cells**, which wrap themselves round and round the axon. This provides electrical insulation around the axon, which greatly speeds up the transmission of action potentials. Not all neurones are myelinated.

Action potentials

Neurones, like all cells, have sodium-potassium pumps in their cell surface membranes. However, in neurones these are especially active. They pump out sodium ions and bring in potassium ions, by active transport. Three sodium ions are moved out of the cell for every two potassium ions that are moved in.

There are also other channels in the membrane that allow the passage of sodium and potassium ions. These are **voltage-gated channels**. The potential difference across the membrane determines whether they are open or closed. When a neurone is resting, quite a few potassium ion channels are open, so potassium ions are able to diffuse back out of the cell, down their concentration gradient.

As a result, the neurone has more positive ions outside it than inside it. This means there is a potential difference (a voltage) across the axon membrane. It has a charge of about **–70mV** (millivolts) inside compared with outside. This is called the **resting potential**.

When a receptor receives a stimulus, this can reduce the potential difference across the membrane, which causes sodium ion channels to open. This allows sodium ions to flood into the cell, down an **electro-chemical gradient**. (The 'electro' gradient refers to the difference in charge across the membrane. The 'chemical' gradient is the difference in concentration of sodium ions.) This quickly reverses the potential difference across the cell membrane, making it much less negative inside. The neurone is said to be **depolarised**. Indeed, the sodium ions keep on flooding in until the cell has actually become positive inside, reaching a potential of about +30mV. The sodium ion channels then close.

This change in potential difference across the membrane causes a set of potassium ion channels to open. Potassium ions can now flood out of the axon, down their electrochemical gradient. This makes the charge inside the axon less positive. It quickly drops back down to a little below the value of the resting potential.

The voltage-gated potassium ion channels then close and the resting potential is restored.

This sequence of events is called an **action potential**. The time taken for the axon to restore its resting potential after an action potential is called the **refractory period**. The axon is unable to generate another action potential until the refractory period is over.

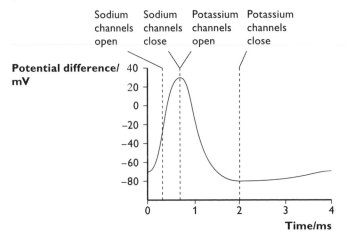

An action potential

Transmission of action potentials

An action potential or nerve impulse that is generated in one part of a neurone travels rapidly along its axon or dendron. This happens because the depolarisation of one part of the membrane sets up local circuits with the areas on either side of it. These cause depolarisation of these regions as well. The nerve impulse therefore sweeps along the axon.

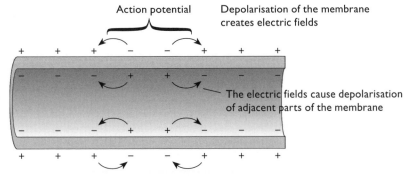

How a nerve impulse travels along a neurone

In a myelinated neurone, local circuits cannot be set up in the parts of the neurone where the myelin sheath is present. Instead, the nerve impulse 'jumps' from one **node of Ranvier** to the next. This is called **saltatory conduction**. This greatly increases the speed at which the action potential travels along the axon.

Synapses

Where two neurones meet, they do not actually touch. There is a small gap between them called a **synaptic cleft**. The membrane of the neurone just before the synapse is called the **presynaptic membrane**, and the one on the other side is the **post-synaptic membrane**. The whole structure is called a **synapse**.

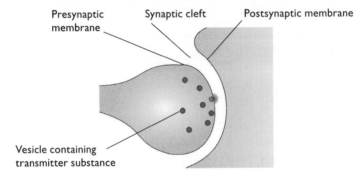

A synapse

- When an action potential arrives at the presynaptic membrane, it causes voltage-gated calcium ion channels to open.
- Calcium ions rapidly diffuse into the cytoplasm of the neurone, down their concentration gradient.
- The calcium ions affect tiny vesicles inside the neurone, which contain a **transmitter substance** such as **acetylcholine**. These vesicles move towards the pre-synaptic membrane and fuse with it, releasing their contents into the cleft.
- The transmitter substance diffuses across the cleft and slots into receptor molecules in the postsynaptic membrane.
- This causes sodium ion channels to open, so sodium ions flood into the cytoplasm of the neurone, depolarising it.
- This depolarisation sets up an action potential in the postsynaptic neurone.

Functions of synapses in the body

- Although action potentials are able to travel in either direction along a neurone, synapses ensure that action potentials can only travel one way.
- One neurone may have synapses with many other neurones. This allows interconnection of nerve pathways from different parts of the body.
- Synapses allow for a wide variety of responses by effectors. For example, a motor neurone may need to receive transmitter substance from many different neurones forming synapses with it before an action potential is generated in it. Or some of these neurones may produce transmitter substances that actually *reduce* the chance of an action potential being produced. The balance between the signals from all these different synapses determines whether or not an action potential

is produced in the motor neurone, and therefore whether or not a particular effector takes action.

The roles of receptors

We have seen that receptors are cells or tissues that sense changes in the internal or external environment. Many types of receptors transform energy from a stimulus into the energy of an action potential in a sensory neurone. The table shows some examples.

Receptor	Type of energy received and transformed to energy in an action potential
Rod and cone cells in the retina	Light
Pacinian corpuscles in the skin	Pressure
Temperature receptors in the skin	Heat
Taste and smell receptors in the tongue and nose	Chemical potential energy in molecules
Hair cells in cochlea in ear	Sound (vibrational) energy

The mammalian endocrine system

The endocrine system is made up of a number of **endocrine glands**. These are organs containing cells that secrete **hormones**. The hormones are secreted directly into the blood, not into a duct as with other types of gland (for example the salivary glands, which secrete saliva into the salivary ducts).

Hormones secreted by an endocrine gland are transported in solution in the blood plasma all over the body. Certain cells have receptors for these hormones in their cell surface membranes. These cells are the **target cells** for the hormones. For example, the target cells for ADH are the cells lining the collecting duct in a nephron.

The control of blood glucose concentration

Blood glucose concentration should remain at a fairly constant value of about 100 mg glucose per 100 cm^3 of blood.

- If blood glucose concentration falls well below this level, the person is said to be hypoglycaemic. Cells do not have enough glucose to carry out respiration, and so metabolic reactions may not be able to take place and the cells cannot function normally. This is especially so for cells such as brain cells, which can only use glucose and not other respiratory substrates. The person may become unconscious and various tissues can be damaged.
- If blood glucose concentration rises well above this level, the person is said to be hyperglycaemic. The high glucose concentration decreases the water potential of the blood and tissue fluid, so that water moves out of cells down a water potential gradient. Again, unconsciousness can result.

Several hormones are involved in the control of blood glucose concentration by **negative feedback**. They include **insulin** and **glucagon**.

Both of these are small proteins. They are secreted by patches of tissue called **islets of Langerhans** in the pancreas. Insulin is secreted by β cells. Glucagon is secreted by α cells.

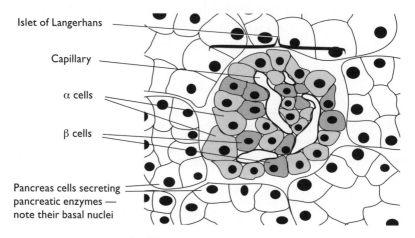

Islet of Langerhans

Capillary

α cells

β cells

Pancreas cells secreting pancreatic enzymes — note their basal nuclei

The histology of the pancreas

When blood glucose concentration rises too high, this is sensed by the β cells. They respond by secreting greater quantities of insulin into the blood. The insulin has several effects, including:

- causing muscle and adipose tissue cells (fat cells) to absorb more glucose from the blood;
- causing liver cells to convert glucose to glycogen for storage.

These effects cause the blood glucose concentration to fall.

When blood glucose concentration falls too low, this is sensed by the α cells. They respond by secreting greater quantities of glucagon into the blood. This has several effects, including:

- causing liver cells to break down glycogen to glucose, and releasing it into the blood;
- causing liver cells to produce glucose from other substances such as amino acids or lipids.

These effects cause blood glucose concentration to rise.

Coordination in plants

As in animals, it is essential that cells in a plant can communicate with each other. This allows them to coordinate their activities appropriately, and allows the entire plant to respond to changes in its internal or external environment. As in animals, plant communication systems involve **receptors** and **effectors**.

Plants do have a system of electrical communication similar to that of an animal's nervous system, but there do not appear to be any truly specialised cells adapted for this function, and 'action potentials' travel only very slowly and are very weak.

Much more is known about **plant hormones**, sometimes called **plant growth substances**. Like animal hormones, these are chemicals that act on target cells, where they bind with receptors either on the cell surface membrane or inside the cell, and bring about changes.

Unlike animal hormones, plant hormones are not made in specialised glands.

Many different plant hormones have been discovered, but there is still much that we do not know about their actions. They often have different effects at different concentrations, in different parts of a plant, at different stages of its life cycle or when other hormones are present.

Auxins

Auxins are a group of plant hormones that are produced by cells in regions of cell division, or **meristems**. In a plant shoot, the apical bud (the bud at the tip of the growing shoot) contains a meristem. Auxin is constantly made here. Auxin molecules are then moved down the shoot from cell to cell, through special auxin transporter proteins in the cell surface membranes.

The action of these transporter proteins in different membranes of different cells causes auxin molecules to accumulate at the **lateral buds** (the buds at the side of the shoot). The auxin inhibits their growth. This is called **apical dominance**.

This means that, as long as there is an actively growing bud at the top of the shoot, the shoot will keep growing upwards and will not branch out sideways. If the apical bud is cut off, then auxin does not accumulate by the lateral buds and they begin to grow, causing the shoot to branch out sideways.

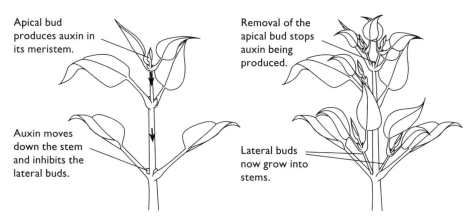

Apical bud produces auxin in its meristem.

Auxin moves down the stem and inhibits the lateral buds.

Removal of the apical bud stops auxin being produced.

Lateral buds now grow into stems.

Auxin and apical dominance

Gibberellins

Stem elongation

Gibberellin, also known as GA, is a hormone that causes stems to elongate. Most plants have genes that cause GA to be produced, but in a plant that lacks these genes the stems stay short and the plant is dwarfed. The application of GA to a dwarf plant makes its stems grow long. Gibberellin works by switching on a gene that is necessary for the growth of the stems.

Germination of cereal seeds

GA is produced in the embryos of germinating seeds, including wheat and barley, after they have absorbed water. GA switches on several genes that encode enzymes that hydrolyse food reserves. These include starch, protein and lipid that are stored in the endosperm of the seed. For example, amylase is produced, which hydrolyses starch to maltose. The soluble maltose is transported to the embryo and used as an energy source and a raw material for the production of new cells.

Abscisic acid

Abscisic acid, ABA, is sometimes known as the plant 'stress hormone'. It is secreted when a plant's environmental conditions become tough, such as very dry, very cold or very hot. For example, when a plant is very short of water, ABA is secreted in the leaves. The guard cells around the stomata respond to the ABA by becoming flaccid and closing the stomata. This response is fast, indicating that ABA, unlike GA, does not work by affecting gene expression.

O Inherited change

Meiosis

A few cells in the human body — some of those in the testes and ovaries — are able to divide by a type of division called **meiosis**. Meiosis involves two divisions (not one as in mitosis), so four daughter cells are formed. Meiosis produces four new cells with:

- only half the number of chromosomes as the parent cell
- different combinations of alleles from each other and from the parent cell

In animals and flowering plants, meiosis produces **gametes**.

You have seen that, in a human, body cells are diploid, containing two complete sets of chromosomes. Meiosis produces gametes that are haploid, containing one complete set of chromosomes.

Before meiosis begins, DNA replication takes place exactly as it does before mitosis (page 52). However, in the early stages of meiosis **homologous chromosomes** (the two 'matching' chromosomes in a nucleus) pair up. The diagram on page 173 shows how meiosis takes place.

Prophase I

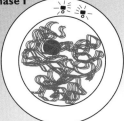 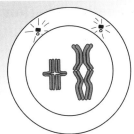

- Centrioles divide and are involved in spindle formation.
- Chromosomes beginning to condense.

- The spindle continues to form.
- Chromosomes pair and are now visible as bivalents.

Metaphase I

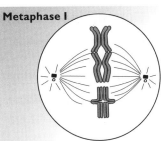

- The nuclear envelope disappears.
- The bivalents are arranged on the equator.

Anaphase I

- Homologous chromosomes separate to opposite poles, pulled by spindle fibres.

Telophase I and cytokinesis

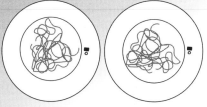

- Chromosomes partly unwind
- Spindle disappears and the nuclear envelopes form
- The cell divides

Prophase II

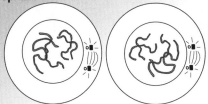

- Centrioles divide and spindles form
- Chromosomes condense

Metaphase II

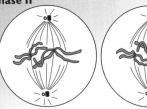

- Individual chromosomes line up on equator.

Anaphase II

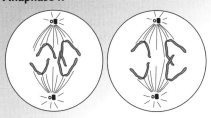

- Centromeres split
- Chromatids move apart, pulled by spindle fibres

Telophase II

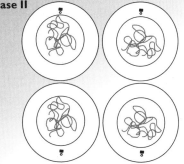

- Spindles disappear
- Nuclear envelopes appear
- Chromosomes unwind

Meiosis

How meiosis causes genetic variation

Meiosis produces genetic variation. This is done by
- crossing over
- independent assortment.

Further genetic variation is produced when gametes produced by meiosis fuse together to produce a zygote.

Crossing over

Each chromosome in a homologous pair carries genes for the same characteristics at the same locus. The alleles of the genes on the two chromosomes may be the same or different.

During prophse of meiosis I, as the two homologous chromosomes lie side by side, their chromatids form links called **chiasmata** (singular: chiasma) with each other. When they move apart, a piece of chromatid from one chromosome may swap places with a piece from the other chromosome. This is called **crossing over**. It results in each chromosome having different combinations of alleles than it did before.

Homologous pair of chromosomes  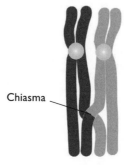 Pieces of each chromatid have swapped places, resulting in new combinations of alleles

Chiasma

Crossing over in meiosis

Independent assortment

Another feature of meiosis that results in the shuffling of alleles — and therefore genetic variation — is **independent assortment**. During the first division of meiosis, the pairs of homologous chromosomes line up on the equator before being pulled to opposite ends of the cell. Each pair behaves independently from every other pair, so there are many different combinations that can end up together. The diagram shows the different combinations you can get with just two pairs of chromosomes. In a human there are 23 pairs, so there is a huge number of different possibilities.

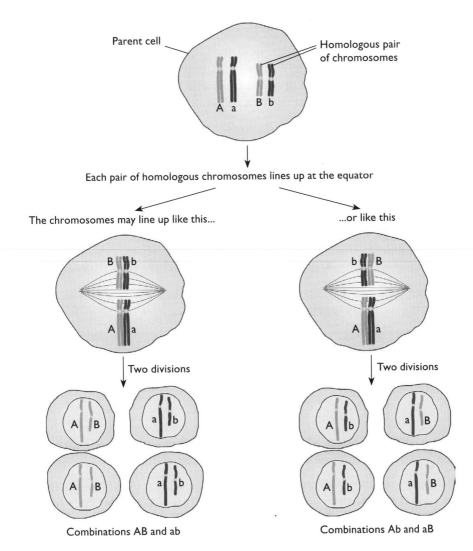

Independent assortment in meiosis

Genetics

Genes are passed from parents to offspring inside the nuclei of gametes. During sexual reproduction, gametes fuse to produce a zygote, which contains one set of chromosomes from each parent. The study of the passage of genes from parent to offspring is called genetics.

Genes and alleles

The two chromosomes in a diploid cell that are similar (e.g. the two chromosome 1s) are said to be **homologous**. They each contain the same genes in the same position, known as the **locus** of that gene. This means that there are two copies of each gene in a diploid cell.

Genes often come in different forms. For example, the gene for a protein that forms a chloride transporter channel in cell surface membranes, called the CFTR protein, has a normal form and several different mutant forms. These different forms of a gene are called **alleles**.

Homozygote and heterozygote

An organism that has two identical alleles for a particular gene is a **homozygote**. An organism that has two different alleles for a particular gene is a **heterozygote**.

Dominant and recessive

We can use letters to represent the different alleles of a gene. For example, we could use F to represent the normal cystic fibrosis allele, and f to represent a mutant allele.

There are three possible combinations of these alleles in a diploid organism: FF, Ff or ff. These are the possible **genotypes** of the organism.

These different genotypes give rise to different **phenotypes** — the observable characteristics of the organism.

genotype	phenotype
FF	normal
Ff	normal
ff	cystic fibrosis

A person with the genotype Ff is said to be a **carrier** for cystic fibrosis, because they have the cystic fibrosis allele but do not have the condition.

The Ff genotype does not cause cystic fibrosis because the F allele is **dominant** and the f allele is **recessive**. A dominant allele is one that is expressed (has an effect) in a heterozygous organism. A recessive allele is one that is only expressed when a dominant allele is not present.

- The dominant allele should always be symbolised by a capital letter, and the recessive allele by a small letter. The same letter should be used for both (not F and c, for example).
- If you are able to choose the symbols that you use in a genetics question, then choose ones where the capital and small letter are different in shape, to avoid confusion (not C and c, for example).

Monohybrid inheritance

This is the inheritance of a single gene.

For example, imagine that a man with the genotype Ff and a woman with the genotype FF have children. In their testes and ovaries, gametes are produced. In the man, half of his sperm will contain the F allele and half will contain the f allele. All of the woman's eggs will contain the F allele.

We can predict the likely genotypes of any children that they have using a **genetic diagram**. Genetic diagrams should always be set out like this:

Parents' genotypes   Ff   ×   FF

Gametes' genotypes   (F)   (f)   (F)

Offspring genotypes and phenotypes

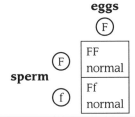

eggs

(F)

(F) sperm	FF normal
(f)	Ff normal

We can therefore predict that there is an equal chance of any child born to them having the genotype FF or Ff. There is no chance they will have a child with cystic fibrosis.

If both parents have the genotype Ff:

Parents' genotypes   Ff   ×   Ff

Gametes' genotypes   (F)   (f)   (F)   (f)

Offspring genotypes and phenotypes

eggs

	(F)	(f)
(F) sperm	FF normal	Ff normal
(f)	Ff normal	ff cystic fibrosis

We can therefore predict that, each time they have a child, there is a 1 in 4 chance that it will have the genotype ff and have cystic fibrosis.

Notice:
- In the examination, it may be important to show the whole genetic diagram, not just the Punnett square (the table showing the genotypes of the offspring).
- The 'gamete genotype' line in the genetic diagram shows the *different kinds* of gametes the parents can produce. If there is only one kind of gamete, you only need to write down one kind - there is no need to write the same one down twice. (If you do that, it won't make any difference to your final answer, but will make twice as much work in writing it all down.)
- The gamete genotypes are shown with a circle drawn around them. This is a convention — if you do this, the examiner will understand that they represent gametes.

- There is often a mark for stating the phenotype produced by each genotype amongst the offspring. The easiest and quickest way to do this is to write the phenotypes in the boxes in the Punnett square.
- The genotypes inside the Punnett square show the *chances* or *probabilities* of each genotype being produced. If you have four genotypes, as in the example above, this does not mean there will be four offspring. It means that, every time an offspring is produced, there is a 1 in 4 chance it will be FF, a 1 in 4 chance it will be ff and a 2 in 4 (better written as 1 in 2) chance it will be Ff.
- An alternative way of writing 'a 1 in 4 chance' is 'a probability of 0.25', or 'a 25% probability'.
- You could also give the final answer in terms of expected ratios. For example, 'We would expect the ratio of unaffected offspring to offspring with cystic fibrosis to be 3:1'.
- The chance of any individual child inheriting a particular genotype is unaffected by the genotype of any previous children. Each time a child is conceived the chances are the same, as shown in the genetic diagram above.

Test crosses

A test cross is a breeding experiment that is carried out to determine the genotype of an organism that shows the dominant characteristic.

For example, in a species of mammal, the gene for hair colour has two alleles, B and b. Allele B gives brown fur and allele b gives white fur.

genotype	phenotype
BB	brown fur
Bb	brown fur
bb	white fur

If an animal has brown fur, we do not know if its genotype is BB or Bb. We can find out by breeding it with an animal with white fur, whose genotype must be bb.

If there are any white offspring, then the unknown animal must have the genotype Bb, as it must have given a b allele to this offspring.

If there are no white offspring, then the unknown animal probably has the genotype BB. However, it is still possible that it is Bb and, just by chance, none of its offspring inherited the b allele from it.

Codominance

In the cystic fibrosis example, one allele is dominant and the other recessive. Some alleles, however, are **codominant**. Each allele has an effect in a heterozygote.

For codominant alleles, it is not correct to use a capital and small letter to represent them. Instead, a capital letter is used to represent the gene, and a superscript to represent the allele.

For example, in some breeds of cattle there are two alleles for coat colour:
C^R is the allele for red coat
C^W is the allele for white coat

genotype	phenotype
$C^R C^R$	red coat
$C^R C^W$	coat with a mixture of red and white hairs — roan
$C^W C^W$	white coat

The inheritance of codominant alleles is shown using a genetic diagram just like the one on page 177. You might like to try showing that two roan cattle would be expected to have offspring with roan, red and white coats in the ratio 2:1:1.

Multiple alleles

Many genes have more than two alleles. For example, the gene that determines the antigen on red blood cells, and therefore your blood group, has three alleles, I^A, I^B and I^O.

I^A and I^B are codominant. They are both dominant to I^O, which is recessive.

genotype	phenotype
$I^A I^A$	blood group A
$I^A I^B$	blood group AB
$I^A I^O$	blood group A
$I^B I^B$	blood group B
$I^B I^O$	blood group B
$I^O I^O$	blood group O

Genetic diagrams involving multiple alleles are constructed in the same way as before.

For example, you could be asked to use a genetic diagram to show how parents with blood groups A and B could have a child with blood group O.

Parents' phenotypes   Blood group A   ×   Blood group B

Parents' genotypes $I^A I^O$ $I^B I^O$

Gametes' genotypes (I^A) (I^O) (I^B) (I^O)

Offspring genotypes and phenotypes

	(I^A)	(I^O)
(I^B)	$I^A I^B$ blood group AB	$I^B I^O$ blood group B
(I^O)	$I^A I^O$ blood group A	$I^O I^O$ blood group O

When solving a problem like this:
- Begin by working out and then writing down a table of genotypes and phenotypes, which you can easily refer back to you as you work through the problem.

- Consider whether you can tell the genotype of any of the individuals in the problem from their phenotype. Here, we know that the child with blood group O must have the genotype $I^o I^o$.
- Work back from there to determine the genotypes of the parents. In this case, each of them must have had an I^o allele to give to this child.
- Always show a complete genetic diagram. Do not take short cuts, even if you can see the answer straight away, as there will be marks for showing each of the steps in the diagram.

Sex linkage

In a human cell, there are two sex chromosomes. A woman has two X chromosomes. A man has an X chromosome and a Y chromosome.

The X chromosome is longer than the Y chromosome. Most of the genes on the X chromosome are not present on the Y chromosome. These are called sex-linked genes.

For example, a gene that determines the production of red-receptive and green-receptive pigments in the retina of the eye is found on the X chromosome. There is a recessive allele of this gene that results in red-green colour-blindness.

A woman has two copies of this gene, because she has two X chromosomes. A man has only one copy, because he has only one X chromosome.

If the normal allele is A, and the recessive abnormal allele is a, then these are the possible genotypes and phenotypes a person may have:

genotype	phenotype
$X^A X^A$	female with normal vision
$X^A X^a$	female with normal vision
$X^a X^a$	female with colour blindness
$X^A Y$	male with normal vision
$X^a Y$	male with colour blindness

Notice:
- Sex-linked genes are shown by writing the symbol for the allele as a superscript above the symbol for the X chromosome.
- There are 3 possible genotypes for a female, but only 2 possible genotypes for a male.

For example, you could be asked to predict the chance of a woman who is a carrier for colour blindness (that is, heterozygous) and a man with normal vision having a colour-blind child.

Parents' phenotypes	woman with normal vision	× man with normal vision
Parents' genotypes	$X^A X^a$	$X^A Y$
Gametes' genotypes	$\boxed{X^A}$ $\boxed{X^a}$	$\boxed{X^A}$ $\boxed{Y}$

Offspring genotypes and phenotypes

	X^A	X^a
X^A	$X^A X^A$ female with normal vision	$X^A X^a$ female with normal vision
Y	$X^A Y$ male with normal vision	$X^a Y$ colour-blind male

There is therefore a 1 in 4 chance that a child born to this couple will be a colour-blind boy. There is a 1 in 2 chance of any boy that is born being colour-blind.

Notice:

- Always show the X and Y chromosomes when working with sex-linked genes.
- A boy cannot inherit a sex-linked gene from his father, because he only gets a Y chromosome from his father. His X chromosome (which carries sex-linked genes) comes from his mother.

Dihybrid inheritance

This involves the inheritance of two genes.

For example, a breed of dog may have genes for hair colour and leg length.

Allele A is dominant and gives brown hair. Allele a is recessive and gives black hair.

Allele L is dominant and gives long legs. Allele l is recessive and gives short legs.

As before, always begin by writing down all the possible genotypes and phenotypes.

genotype	phenotype
AALL	brown hair, long legs
AaLL	brown hair, long legs
aaLL	black hair, long legs
AALl	brown hair, long legs
AaLl	brown hair, long legs
aaLl	black hair, long legs
AAll	brown hair, short legs
Aall	brown hair, short legs
aall	black hair, short legs

Notice:

- Always write the two alleles of one gene together. Do not mix up alleles of the two different genes.
- Always write the alleles in the same order. Here, we have decided to write the alleles for hair colour first, followed by the alleles for leg colour. Don't swap these round part way through.

A dihybrid cross

A genetic diagram showing a dihybrid cross is set out exactly as for a monohybrid cross.

The only difference is that there will be more different types of gametes.

Each gamete will contain just one allele of each gene.

This genetic diagram shows the offspring we would expect from a cross between two dogs that are both heterozygous for both genes.

| Parents' phenotypes | brown hair, long legs | × | brown hair, long legs |

Parents' genotypes — AaLl — AaLl

Gametes' genotypes — (AL) (Al) (aL) (al) — (AL) (Al) (aL) (al)

Offspring genotypes and phenotypes

	(AL)	(Al)	(aL)	(al)
(AL)	AALL brown, long	AALl brown, long	AaLL brown, long	AaLl brown, long
(Al)	AALl brown, long	AAll brown, short	AaLl brown, long	Aall brown, short
(aL)	AaLL brown, long	AaLl brown, long	aaLL black, long	aaLl black, long
(al)	AaLl brown, long	Aall brown, short	aaLl black, long	aall black, short

We would therefore expect offspring in the ratio 9 brown hair, long legs : 3 brown hair, short legs : 3 black hair, long legs : 1 black hair, short legs.

Notice:

- Each gamete contains one allele of each gene (one of either A or a, and one of either L or l).
- The ratio 9:3:3:1 is typical of a dihybrid cross between two heterozygotes, where the alleles of both genes show dominance (as opposed to codominance).

You might like to try an example for yourself to show that, when a homozygous recessive individual is crossed with one that is heterozygous for both genes, the expected ratio would be 1:1:1:1.

Using statistics in genetics

Genetic diagrams can be used to tell us the probabilities of particular genotypes occurring in offspring in particular ratios. However, because these are just *probabilities*, the actual phenotypic ratios are rarely exactly as we predicted.

For example, imagine that a plant has genes for flower colour (allele Y for yellow and y for white flowers) and petal size (P for large petals and p for small petals).

We cross two plants that we think are heterozygous for both alleles. We would expect to get offspring with phenotypes:

yellow, large		yellow, small		white, large		white, small
9	:	3	:	3	:	1

However, what we actually get are 847 offspring with these phenotypes:

yellow, large	yellow, small	white, large	white, small
489	151	159	48

What we would *expect* to have got is:

yellow, large	yellow, small	white, large	white, small
476	159	159	53

The question we need to ask is: Are these results close enough to the expected results to indicate that we are right in thinking that the two parents are really heterozygous, and the two genes really are behaving in the way we predicted? Or are they so different from the results we expected that they indicate that something different is happening? In other words, is the difference between our observed and expected results **significant**?

We can use statistics to tell us how likely it is that the difference between our observed results and our expected results is just due to chance, or whether it is so different that we must have been wrong in our predictions.

Null hypothesis and probability level

A statistics test begins by setting up a **null hypothesis**. We then use the statistics test to determine the probability of the null hypothesis being true.

In this case, our null hypothesis would be:

the observed results are not significantly different from the expected results.

We then work through a statistics test, which gives us a *probability of our null hypothesis being correct*. In biology, it is conventional to say that:
- if the probability of the null hypothesis being correct is *greater than or equal to 0.05*, then we can *accept* the null hypothesis.
- if the probability is *less than 0.05* that the null hypothesis is correct, then we must *reject* it.

The chi-squared test

To test the possibility of our null hypothesis being correct, the most suitable statistical test for this set of results is the **chi-squared test**. This can also be written: χ^2 **test**.
- Construct a table similar to the one overleaf. You need one column for each category of your results.

	Yellow, large	Yellow, small	White, large	White, small
Observed numbers, O				
Expected numbers, E				
O–E				
(O–E)²				
$\dfrac{(O-E)^2}{E}$				
$\Sigma\ \dfrac{(O-E)^2}{E}$				

- Fill in the observed numbers and the expected numbers for each category.
- Calculate O–E for each category.
- Calculate (O–E)² for each category.
- Calculate $\dfrac{(O-E)^2}{E}$ for each category.
- Add together all of the $\dfrac{(O-E)^2}{E}$ values, to give you $\Sigma\ \dfrac{(O-E)^2}{E}$

Your table now looks like this.

	Yellow, large	Yellow, small	White, large	White, small
Observed numbers, O	489	151	159	48
Expected numbers, E	476	159	159	53
O–E	13	–8	0	–5
(O–E)²	169	64	0	25
$\dfrac{(O-E)^2}{E}$	0.35	0.42	0	0.47
$\Sigma\ \dfrac{(O-E)^2}{E} = 1.24$				

The number 1.24 is the chi-squared value.

You now need to look this up in a probability table, to find out what it tells us about the probability of the null hypothesis being correct.

This is a part of a probability table for chi-squared values.

Degrees of freedom	Probability of null hypothesis being correct			
	0.1	0.05	0.01	0.001
1	2.71	3.84	6.64	10.83
2	4.60	5.99	9.21	13.82
3	6.25	7.82	11.34	16.27
4	7.78	9.49	13.28	18.46

- The numbers inside the cells in the table are chi-squared values.
- The numbers in the first column are degrees of freedom. You have to choose the correct row in the table for the number of degrees of freedom in your data. In general:

degrees of freedom = number of different categories – 1

In this case, there were four categories (the four different phenotypes), so there are 3 degrees of freedom.

- Look along the row for 3 degrees of freedom until you find the number closest to the chi-squared value you have calculated. This was 1.24, and all the numbers in the row are much bigger than this. The closest is 6.25.
- Look at the probability associated with this number. It is 0.1. This means that if the chi-squared value was 6.25, there is a 0.1 probability that the null hypothesis is correct.
- Remember that, if the probability of the null hypothesis being correct is equal to or greater than 0.05, you can accept it as being correct. 0.1 is much bigger than 0.05, so you can definitely accept the null hypothesis as being correct. Indeed, because your value of chi-squared was actually much smaller than 6.25, we would need to go a long way further left in the table, which would take us into probabilities even greater than 0.1.
- We can therefore say that the difference between the observed results and expected results is *not significant*. The differences between them are just due to random chance.

Mutation

A mutation is a random, unpredictable change in the DNA in a cell. It may be:
- a change in the sequence of bases in one part of a DNA molecule, or
- an addition of extra DNA to a chromosome, or a loss of DNA from it, or
- a change in the total number of chromosomes in a cell.

Mutations are most likely to occur during DNA replication, for example when a 'wrong' base may slot into position in the new strand being built. Almost all of these mistakes are immediately repaired by enzymes, but some may persist.

A change in the sequence of bases in DNA may result in a change in the sequence of amino acids in a protein. (Note that this does not always happen, because there is more than one triplet that codes for each amino acid, so a change in a triplet may not change the amino acid that is coded for.) This in turn may result in a change in the 3-D structure of the protein and therefore the way that it behaves.

Cystic fibrosis is a genetic condition resulting from a mutation in a gene that codes for a carrier protein called CFTR. You can find more about this condition on pages 202 to 203.

Sickle cell anaemia is a genetic condition resulting from a substitution in the gene coding for the β polypeptide in a haemoglobin molecule. You can find more about this on page 61.

The effect of environment on phenotype

Although genes have major effects on an organism's phenotype, the organism's environment can also have large effects.

Lactase production in *Escherichia coli*

The bacterium *Escherichia coli* has a gene that codes for the production of the enzyme lactase, which hydrolyses the disaccharide lactose to glucose and galactose. This gene is only expressed when the bacterium encounters lactose in its environment.

Hair colour in cats

Many different genes determine hair colour in cats. At least eight different genes, at different loci, are known to influence hair colour and it is thought that there are probably more. These are known as **polygenes**. Depending on the particular combination of alleles that a cat has for each of these genes, it can have any of a very wide range of colours. Hair colour in cats is an example of **continuous variation**. This is variation in which there are no clear-cut categories. There is a continuous range of variation in colour between the very lightest and very darkest extremes.

The cat hair colour genes exert their effect by coding for the production of enzymes. One such gene is found at the *C* locus. Siamese cats have two copies of a recessive allele of this gene called c^s. This gene codes for an enzyme which is sensitive to temperature. It produces dark hair at the extremities of paws, ears and tail where the temperature is lower, and light hair in warmer parts of the body. The colouring of a Siamese cat is therefore the result of interaction between genes and environment.

The hair is darker in areas which are colder, such as the ears and paws

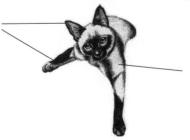

The hair is lighter in areas which are warmer, such as the body

Hair colour in a Siamese cat

Human height

Human height is also affected by many different genes at different loci. It is also affected by environment. Even if a person inherits alleles of these genes that give the potential to grow tall, he or she will not grow tall unless the diet supplies plenty of nutrients to allow this to happen. Poor nutrition, especially in childhood, reduces the maximum height that is attained.

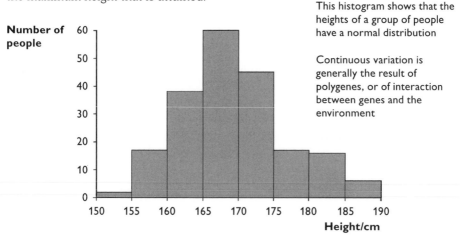

This histogram shows that the heights of a group of people have a normal distribution

Continuous variation is generally the result of polygenes, or of interaction between genes and the environment

Continuous variation in human height

Cancer

The risk of developing cancer is influenced by both genes and environment. For example, a woman with particular alleles of the genes *BRCA1* or *BRCA2* has a 50% to 80% of chance of developing breast cancer at some stage in her life. This is a much higher risk than for people who do not have these alleles. The normal alleles of these genes protect cells from changes that could lead to them becoming cancerous.

However, environment also affects this risk. Smoking, for example, increases the risk even further. Taking the drug tamoxifen can reduce the risk.

Monoamine oxidase A

Monoamine oxidase A (MAO-A) is an enzyme that is found associated with mitochondria in the nervous system, and also in the liver and digestive system. In the nervous system, it is involved in the inactivation of neurotransmitters including noradrenaline and serotonin.

Some alleles of the monoamine oxidase gene produce low activity MAO-A, while others produce high activity MAO-A. It has been found that children with the high activity form, if maltreated, are more likely to show antisocial behaviour than similarly treated children with the low activity form.

Other behaviours, such as novelty seeking, also appear to be associated with particular alleles of this gene. However, in all cases the environment also has large effects on behaviour; behaviour is produced by interaction between this gene (and probably others as yet unidentified) and the environment.

P Selection and evolution

Natural selection

In a population of organisms, each parent can produce more young than are required to maintain the population at a constant level. For example, each pair of adult foxes in a population needs to have only two young in their lifetime to keep the population the same size. However, a pair of foxes may have more than 30 young. The population can **potentially overproduce**. But the population generally stays approximately the same size because most of the young do not survive long enough to reproduce.

In a population, not every organism has exactly the same alleles or exactly the same features. There will be **genetic variation** within the population. Those organisms whose particular set of features are best suited to the environment are most likely to survive. Those with less useful features are more likely to die.

The organisms with the most useful features are therefore more likely to reach adulthood and reproduce. Their alleles will be passed on to their offspring. Over many generations, the alleles that confer useful characteristics on an individual are therefore likely to become more common. Alleles that do not produce such useful characteristics are less likely to be passed on to successive generations and will become less common.

This process is called **natural selection**. Over time, it ensures that the individuals in a population have features that enable them to survive and reproduce in their environment.

Stabilising and directional selection

When the environment is fairly stable, natural selection is unlikely to bring about change. If the organisms in a population are already well adapted to their environment, then the most common alleles in the population will be those that confer an advantage on the organisms, and it is these alleles that will continue to be passed on to successive generations. This is called **stabilising selection**.

However, if the environment changes, alleles that were previously advantageous may become disadvantageous. For example, in a snowy environment the individuals in a species of mammal may have white fur that camouflages them against the snow and confers an advantage in escaping predators. If the climate changes so that snow no longer lies on the ground, then animals with white fur may be more likely to be killed than animals with brown fur. Those with brown fur are now most likely to reproduce and pass on their alleles to the next generation. Over time, brown may become the most common fur colour in the population. This is an example of **directional selection** or **evolutionary selection**.

In a snowy environment, white animals are more camouflaged, making them less visible to predators, so are at a selective advantage. The white animal is most likely to survive.

In an environment without snow, white animals are very noticeable and more likely to be predated. The white animal is least likely to survive.

Stabilising and directional selection

Directional selection may result in **evolution**. Evolution can be defined as a long-term change in the characteristics of a species, or in the frequency of particular alleles within the species.

Sickle cell anaemia and malaria

Similarities between the global distribution of the genetic disease sickle cell anaemia, and the infectious disease malaria, are a result of natural selection.

Sickle cell anaemia is caused by an allele of a gene coding for the β polypeptide in haemoglobin. This is described on page 61.

Malaria is caused by a protoctist, *Plasmodium*, which is transmitted by *Anopheles* mosquitoes. This is described on pages 81–82.

A person who is homozygous for the faulty allele for the β polypeptide, Hb^SHb^S, has sickle cell anaemia. This is a serious disease and a child with this genotype is unlikely to survive to adulthood and have children.

A person who is homozygous for the normal allele for the β polypeptide, Hb^AHb^A, is vulnerable to malaria. In parts of the world where the *Anopheles* mosquitoes that transmit malaria are found, and where malaria is common, a child with this genotype may not survive to adulthood and have children.

A person who is heterozygous for these alleles, Hb^AHb^S, does not have sickle cell anaemia. They are also much less likely to suffer from a serious attack of malaria than a person with the genotype Hb^AHb^A. Therefore a heterozygous person is the most likely to survive to adulthood and have children.

The allele Hb^S therefore continues to survive in populations where malaria is present, because it is passed on from generation to generation by heterozygous people.

In parts of the world where malaria is not common, there is no selective advantage in having this allele, and it is very rare.

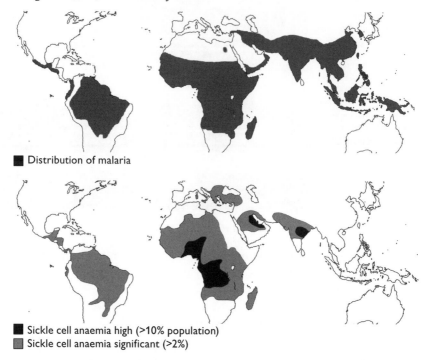

■ Distribution of malaria

■ Sickle cell anaemia high (>10% population)
■ Sickle cell anaemia significant (>2%)

The global distribution of malaria and sickle cell anaemia

Speciation

A **species** is often defined as a group of organisms with similar morphological and physiological characteristics, which are able to breed with each other to produce fertile offspring. So, for example, lions and tigers are distinct species, even though in a zoo they may be persuaded to breed together. Such interbreeding between the two species never occurs in the wild and, in any case, the offspring are not able to breed themselves.

So how are new species produced? We have seen that natural selection can produce changes in allele frequency in a species, but how much change is needed before we can say that a new species has been formed?

The crucial event that must occur is that one population must become unable to interbreed with another. They must become **reproductively isolated** from one another. Once this has happened, we can say that the two populations are now different species.

There are many ways in which reproductive isolation can happen. One which we think has been especially important in the formation of new species of plants and animals begins by a group of individuals in the population becoming **geographically separated** from the rest.

For example, a few lizards might get carried out to sea on a floating log, and be carried to an island where that species of lizard was not previously found. The lizards on this island are subjected to different environmental conditions from the rest of the species left behind on the mainland. Different alleles are therefore selected for in the two groups. Over time, the allele frequency in the lizards on the island becomes very different from the allele frequency in the original, mainland lizards. This may cause their characteristics to become so different that — even if a bridge appears between the island and the mainland — they can no longer interbreed to produce fertile offspring. Reasons for this could include:

- they have evolved different courtship behaviours, so that mating no longer occurs between them.
- the sperm of one group are no longer able to survive in the bodies of the females of the other group, so fertilisation does not occur.
- the number or structure of the chromosomes is different, so that the zygote that is formed by fertilisation does not have a complete double set of genes and cannot develop.
- even if a zygote is successfully produced, the resulting offspring may not be able to form gametes, because its two sets of chromosomes (one from each parent) are unable to pair up with each other successfully and so cannot complete meiosis.

Artificial selection

Artificial selection is the choice, by humans, of which animals or plants are allowed to breed.

For example, farmers may want to have wheat plants that produce large yields of grain and that are resistant to fungal disease such as rust. They want the plants to be short, so that more energy can be put into growing grain (seed) rather than wasted on stems, and also so that the plants are less likely to fall over in heavy rain or high wind.

Breeders can choose a wheat plant that is short, with high yields of grain, and another that does not have these characteristics but is very resistant to rust. They take pollen from one plant and place it on the stigmas of the other. (The anthers of the second plant are first removed, so that it cannot pollinate itself.)

The resulting seeds are collected and sown. The young plants are allowed to grow to adulthood. Then the plants that show the best combination of desired characteristics are bred together.

This continues for several generations, until the breeder has a population of plants that have high yield and high resistance to rust.

Notice:

- Artificial selection is similar to natural selection, in that individuals with particular characteristics are more likely to breed than others.
- However, in artificial selection, individuals without these characteristics will not breed at all, whereas in natural selection there is still a chance that they might breed.
- Artificial selection may therefore produce bigger changes in fewer generations than natural selection normally does.
- The breeders do not need to know anything about the genes or alleles that confer the characteristics they want their plants to have; they just choose the plants with those characteristics and hope that the appropriate alleles will be passed on to the next generation.
- Artificial selection usually requires many generations of selection before the desired result is obtained.

Q Biodiversity and conservation

Classification

Biologists classify organisms according to how closely they believe they are related to one another. Each species has evolved from a previously existing species. We do not usually have any information about these ancestral species, so we judge the degree of relatedness between two organisms by looking carefully at their physiology, anatomy and biochemistry. The greater the similarities, the more closely they are thought to be related.

The system used for classification is a **taxonomic system**. This involves placing organisms in a series of taxonomic units which form a hierarchy. The largest unit is the **kingdom**. Kingdoms are subdivided into phyla, classes, orders, families, genera and species.

An example of classification

Kingdom	Animalia
Phylum	Chordata
Class	Mammalia
Order	Rodentia
Family	Muridae
Genus	*Mus*
Species	*Mus musculus* (house mouse)

The five-kingdom classification system

One method of classification is to place all living organisms into five kingdoms. These are:

Kingdom Prokaryota These are organisms with prokaryotic cells (page 21). This kingdom includes bacteria and blue-green algae.

Kingdom Protoctista These organisms have eukaryotic cells. They mostly exist as single cells, but some are made of groups of similar cells.

Kingdom Fungi Fungi have eukaryotic cells surrounded by a cell wall, but this is not made of cellulose and fungi never have chloroplasts.

Kingdom Plantae These are the plants. They have eukaryotic cells surrounded by cellulose cell walls and they feed by photosynthesis.

Kingdom Animalia These are the animals. They have eukaryotic cells with no cell wall.

Biodiversity

Biodiversity can be defined as the range of habitats, communities and species in an area, and the genetic variation that exists within the populations of each species. (A **community** can be defined as all the different organisms, of all the different species, that live in the same place at the same time.)

Aspects of biodiversity

We can find out something about the biodiversity of an area by measuring the **species richness**. This is the number of different species in the area. The greater the species richness, the greater the biodiversity.

We can also investigate the number of different alleles in the **gene pool** of the species. This is a measure of **genetic diversity**. The gene pool is all the different alleles, of all the different genes, in a population. A **population** can be defined as all the individuals of a particular species that live in the same place at the same time.

It is not possible to discover and count every single allele in a population. Sometimes, researchers may simply record the range of different features in the population, such as the range of hair colour. More usefully, they may collect DNA samples and analyse the base sequences to look for variations between individuals. The more variation in base sequences, the greater the genetic diversity.

It is generally considered desirable for a species to have reasonably large genetic diversity. This means that if the environment changes — for example, because of climate change or if a new pathogen emerges — then at least some of the population may possess features that will enable them to survive. Genetic diversity allows species to become adapted to a changing environment.

Conservation

Conservation aims to maintain biodiversity.

Reasons for maintaining biodiversity
- **Stability of ecosystems** The loss of one of more species within a community may have negative effects on others, so that eventually an entire ecosystem becomes seriously depleted.
- **Benefits to humans** The loss of a species could prevent an ecosystem from being able to supply humans with their needs.
    - For example, an ancient civilisation in Peru, the Nazca people, are thought to have cut down so many huarango trees that their environment became too dry and barren for them to grow crops.
    - Today we make use of many different plants to supply us with drugs. There may be many more species of plants, for example in rainforests, that could potentially provide life-saving medicines.
    - In some countries, such as Kenya, people derive income from tourists who visit the country to see wildlife.
- **Moral and ethical reasons** Many people feel that it is clearly wrong to cause extinction of a species, and that we have a responsibility to maintain environments in which all the different species on Earth can live.

Endangered species
A species is said to be endangered if its numbers have fallen so low that it may not be able to maintain its population for much longer.

For example, tigers are seriously endangered. This has happened for many reasons.

- Tigers are hunted by humans for various body parts that are thought to be effective medicines by people in countries such as China and Korea.
- Tigers require large areas of land to live and hunt successfully. Expanding human populations also require land, and this has greatly reduced the area of suitable habitat for tigers.
- Tigers are large predators that can pose threats to humans and their livestock, and so may be killed.
- The tiger populations in many areas are now so low that genetic diversity is small. This makes it more likely that an individual tiger will inherit the same recessive allele from both parents, which may produce harmful characteristics. It also makes it less likely that the population will be able to adapt (through natural selection acting on natural variation) to changes in its environment.

Methods of protecting endangered species
Habitat conservation
The best way to conserve threatened species is in their own natural habitat. For example, National Parks and nature reserves can set aside areas of land in which the

species is protected. The species will only survive if the features of its habitat that are essential to its way of life are maintained. Maintaining habitat is also likely to be beneficial to many other species in the community.

However, this is often difficult, because people living in that area have their own needs. In parts of the world where people are already struggling to survive, it is difficult to impose restrictions on the way they can use the land where they live. The best conservation programmes involve the local people in habitat conservation and reward them for it in some way. For example, they could be given employment in a National Park, or could be paid for keeping a forest in good condition.

Zoos
- **Captive breeding programmes** This involves collecting together a small group of organisms of a threatened species and encouraging them to breed together. In this way, extinction can be prevented. The breeding programme will try to maintain or even increase genetic diversity in the population by breeding unrelated animals together. This can be done by moving males from one zoo to another, or by using *in vitro* fertilisation using frozen sperm transported from males in another zoo. It may also be possible to implant embryos into a surrogate mother of a different species, so that many young can be produced even if there is only a small number of females of the endangered species.
- **Reintroduction programmes** The best captive breeding programmes work towards reintroducing individuals to their original habitat, if this can be made safe for them. It is very important that work is done on the ground to prepare the habitat for the eventual reintroduction of the animals. For example, the scimitar-horned oryx has been successfully reintroduced to Tunisia, following a widespread captive breeding programme in European zoos and the preparation and protection of suitable habitat, including the education and involvement of people living in or around the proposed reintroduction area.
- **Education** Zoos can bring conservation issues to the attention of large numbers of people, who may decide to contribute financially towards conservation efforts or to campaign for them. Entrance fees and donations can be used to fund conservation programmes both in the zoo itself and in natural habitats.
- **Research** Animals in zoos can be studied to find out more about their needs in terms of food, breeding places and so on. This can help to inform people working on conservation in natural habitats.

Botanic gardens
Botanic gardens are similar to zoos, but for plants rather than animals. Like zoos, they can be safe havens for threatened plant species, and are involved in breeding programmes, reintroduction programmes, education and research.

Seed banks
Seed banks store seeds collected from plants. Many seeds will live for a very long time in dry conditions, but others need more specialised storage environments. A few of the seeds are germinated every so often so that fresh seed can be collected and stored.

Seed banks can help conservation of plants just as zoos can help conservation of animals. The Royal Botanic Gardens, Kew, has a huge seed bank at Wakehurst Place, Sussex, UK. Collectors search for seeds, especially those of rare or threatened species, and bring them to the seed bank where they are carefully stored. Another seed bank, built into the permafrost (permanently frozen ground) in Norway, aims to preserve seeds from all the world's food crops.

R Gene technology

Gene technology is the manipulation of genes in living organisms. Genes from one organism may be inserted into another. This may be done within the same species (for example, in gene therapy) or genes may be transferred from one species to another.

Gene technology for insulin production

Insulin is a small protein. It is a hormone secreted by β cells in the islets of Langerhans in the pancreas in response to raised blood glucose concentration (pages 169–170).

In type I diabetes, no or insufficient insulin is secreted, and the person has to inject insulin. This used to be obtained from animals such as pigs. Today, almost all insulin used in this way is obtained from genetically modified bacteria.

Identifying the insulin gene

The amino acid sequence of insulin was already known. From this, the probable base sequence of the gene that codes for it, and of the mRNA transcribed from the gene, could be worked out.

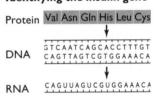

Identifying the insulin gene

Protein | Val Asn Gln His Leu Cys

DNA
GTCAATCAGCACCTTTGT
CAGTTAGTCGTGGAAACA

RNA
CAGUUAGUCGUGGAAACA

Making the human insulin gene

Messenger RNA was extracted from β cells. These cells express the gene for insulin, so much of this mRNA had been transcribed from this gene. The appropriate mRNA was then incubated with the enzyme **reverse transcriptase**, which built single-stranded cDNA molecules against it. These were then converted to double-stranded DNA — the insulin gene.

Some extra single-stranded DNA was then added to each end of the DNA molecules. These are called **sticky ends**. Because they are single-stranded, they are able to form hydrogen bonds with other single-stranded DNA, enabling DNA molecules to join up with one another. This is important in a later stage of the process.

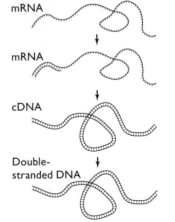

Making the human insulin gene

mRNA

mRNA

cDNA

Double-stranded DNA

Cloning the DNA

Multiple copies of the DNA were then made using **DNA polymerase**. This can be done using the **polymerase chain reaction**, or **PCR**. A small amount of DNA is incubated with DNA polymerase in a repeated sequence of changing temperatures, enabling a huge number of copies to be made in a relatively short period of time.

Cloning the DNA

Inserting the DNA into a plasmid vector

A **plasmid** is a small, circular DNA molecule found in many bacteria. A plasmid was cut open using a **restriction enzyme**. (These are enzymes that occur naturally in bacteria, where they attack and destroy viral DNA that may enter the bacterium.) The restriction enzymes make a stepped cut across the DNA molecule, leaving single-stranded regions.

The cut plasmids and the insulin gene (the cloned DNA) were then mixed together, along with the enzyme **DNA ligase**. Complementary base pairing took place between the sticky ends added to the insulin genes and the sticky ends of the cut plasmids. DNA ligase then joined up the sugar-phosphate backbones of the DNA strands. This resulted in closed plasmids containing the insulin gene.

Genes conferring resistance to an antibiotic were also introduced into the plasmids, next to the insulin gene.

Not all of the plasmids took up the gene. Some just closed back up again without it.

Cutting a plasmid with a restriction enzyme

Inserting the insulin gene into a plasmid

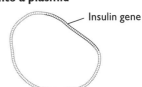
Insulin gene

Inserting the plasmid vector into a bacterium

The plasmids were mixed with a culture of the bacterium *Escherichia coli*. About 1% of them took up the plasmids containing the insulin gene.

Inserting the plasmid vector into a bacterium

E. coli

Identifying the genetically modified bacteria

Antibiotics were then added to the culture of *E. coli* bacteria. The only ones that survived were the ones that had successfully taken up the plasmids containing the antibiotic resistance gene. Most of these plasmids would also have contained the insulin gene. Most of the surviving *E. coli* bacteria were therefore ones that now contained the human insulin gene.

Cloning the bacteria and harvesting the insulin

The bacteria were then grown in fermenters, where they were provided with nutrients and oxygen to allow them to reproduce to form large populations. Reproduction is asexual, so all the bacteria were genetically identical (clones).

This is now done on a large scale. The bacteria synthesise and secrete insulin, which is harvested from the fermenters and purified before sale.

Identifying the modified bacteria

Bacteria cultured on agar + antibiotic

Surviving bacteria

Growing the transformed bacteria

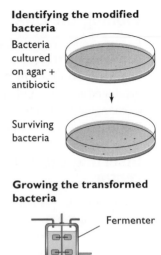

Fermenter

Advantages of insulin produced by gene technology

- The insulin produced by the genetically engineered *E. coli* is identical to human insulin, because it is made following the genetic code on the human insulin gene. Insulin obtained from the pancreas of an animal is slightly different, and therefore may have different effects when used to treat diabetes in humans.
- Large quantities of insulin can be made continuously using *E. coli*, and this can be done under controlled conditions. Only small quantities of insulin can be obtained from the pancreas of an animal, and it is not easy to purify the insulin to produce a standard product that is safe for medicinal use.
- Many religions and cultures, and also many individuals, are uncomfortable with the idea of harvesting insulin from a dead animal for use in humans.

Promoters

In bacteria, each gene is associated with a region of DNA called a **promoter**. The enzyme **RNA polymerase** must bind to the promoter before it can begin transcribing the DNA to produce mRNA.

It is therefore important to ensure that there is a promoter associated with the human insulin gene when it is inserted into *E. coli*.

Markers

The antibiotic resistance genes added to the plasmids along with the human insulin gene act as **markers**. They make it possible to identify the bacteria that have taken up the gene.

There is concern that using antibiotic resistance genes as markers could increase the likelihood of the development of populations of harmful bacteria that are resistant to antibiotics. Today, the most common markers used are genes that code for the production of **fluorescent green protein**. The gene for this protein can be inserted along with the desired gene. Cells that fluoresce green are therefore likely to have taken up the desired gene.

Issues relating to gene technology

The ability to manipulate genes has opened up new possibilities that had not previously been foreseen. While some new technologies have obvious advantages and have wide public acceptance, others have been less well received.

Many of the arguments against gene technology undoubtedly arise from ignorance. For example, in a recent survey into people's views about using GM (genetically modified) tomatoes, 59% believed that ordinary tomatoes did not contain DNA. It is important to look rationally at arguments and not dismiss or accept statements without good knowledge of the underlying science. On the other hand, if people believe passionately that a particular technology is wrong then — even if the science does not support their view — it could be considered wrong to force them to accept it.

You should watch the media for current examples of gene technology that are fuelling debate. You will also find a great deal of information and opinion on the internet. Look carefully to see the source of the material, as this will help you to judge how much trust you can place in it. In general, academic sources can be relied upon to be scientifically sound and unbiased.

The arguments are wide-ranging and very diverse, so it is impossible to list all of them here. Two examples are described below.

Golden Rice™

See also page 218. This is rice that has had genes encoding vitamin A added to it. In countries where rice forms a major part of the diet, children may suffer from vitamin A deficiency, causing blindness. Eating this GM (genetically modified) rice provides more β carotene, which is used to make vitamin A, and so can help to avoid this condition.

The company and researchers who developed the rice have donated it free for use in developing countries. However, some people argue that this is not the best way to solve problems associated with poverty, and that it would be better to improve people's lives so that they are able to choose a wider variety of foods to eat. They are also concerned that the GM rice might somehow be harmful to health.

Bt maize

See also pages 217–218. Maize is a very widely grown cereal crop. Its grain is used as food for humans and for animals. There are numerous insect pests, for example the corn borer, which greatly reduce the yield of maize. Crops can be sprayed with insecticides to kill the corn borers.

A gene from the bacterium *Bacillus thuringiensis* has been introduced to maize to produce Bt maize. This gene encodes a protein that is toxic to corn borers. Bt maize is therefore resistant to these pests. This reduces the amount of insecticide that is sprayed, and therefore reduces harm to other insects.

However, many people do not like the idea of eating maize that contains the Bt toxin, and think it could harm human health. Research has so far uncovered no adverse effects.

There is some evidence that insects other than corn borers may be harmed. For example, research has indicated that butterflies feeding on pollen from the maize may be harmed. Other research indicates that insect larvae in streams running close to fields of Bt maize may be harmed. However, the evidence is not strong and debate continues.

Electrophoresis

Electrophoresis is a way of separating strands of DNA of different lengths.

Before electrophoresis is carried out, a sample of DNA is exposed to a set of **restriction enzymes**. These enzymes cut DNA molecules where particular base sequences are present. For example, a restriction enzyme called *Bam*H1 cuts where the base sequence -GGATCC- is present on one strand of the DNA. Other restriction enzymes target different base sequences. This cuts the DNA into fragments of different lengths.

To carry out gel electrophoresis, a small, shallow tank is partly filled with a layer of agarose gel. A potential difference is applied across the gel, so that a direct current flows through it.

A mix of the DNA fragments to be separated is placed on the gel. DNA fragments carry a small negative charge, so they slowly move towards the positive terminal. The larger they are, the more slowly they move. After some time, the current is switched off and the DNA fragments stop moving through the gel.

The DNA fragments must be made visible in some way, so that their final positions can be determined. This can be done by adding fluorescent markers to the fragments. Alternatively, single strands of DNA made using radioactive isotopes, and with base sequences thought to be similar to those in the DNA fragments, can be added to the gel. They will pair up with fragments which have complementary base sequences, so their positions are now emitting radiation. This can be detected by its effects on a photographic plate.

Using electrophoresis in genetic fingerprinting

Some regions of human DNA are very variable, containing variable numbers of repeated DNA sequences. These are known as variable number tandem repeats, or **VNTR**s. Each person's set of VNTR sequences is unique. Only identical twins share identical VNTR sequences.

Restriction enzymes are used to cut a DNA sample near VNTR regions. The chopped pieces of DNA are then separated using gel electrophoresis. Long VNTR sequences do not travel as far on the gel as short ones. The pattern of stripes produced is therefore determined by the particular combination of VNTRs that a person has.

This can be used to:

- determine if a sample of semen, blood or other tissue found at a crime scene could have come from the victim or a suspect;
- determine whether a particular person could be the child, mother or father of another.

The dark areas on these autoradiographs of DNA represent particular DNA sequences.

In the first profile, you can see that all bands on the child's results match up with either the mother's or potential father's, so this man could be the child's father.

However, on the second one, the child has a band that is not present in either the mother's or the father's results. Some other person must therefore be the child's father.

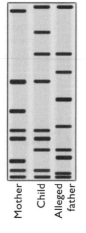

Electrophoresis in paternity testing

Using electrophoresis in DNA sequencing

DNA sequencing is the determination of the base sequence in a length of DNA.

There are many different methods of DNA sequencing, a number of which are now fully automated. One of these is called the **chain-terminator method**.

- Multiple copies of the single-stranded DNA to be sequenced are made, using the PCR reaction.
- The copies are divided into four sets. Into one set, normal DNA nucleotides containing A, C and G are added. However, the nucleotide containing T is modified so that it is unable to join to the next nucleotide. This means that whenever a T nucleotide is added to a growing chain, the chain stops.

In the other three sets, the modified nucleotide is one of A, C or G.

- DNA polymerase is added to all four mixtures, allowing DNA synthesis to take place, using the single-stranded DNA as a template.
- The new DNA fragments are then separated using gel electrophoresis. Each set is run on a different lane in the gel. Each bar on the gel represents a DNA fragment ending with a particular base. This can be used to work out the sequence of the bases along the length of the original DNA.

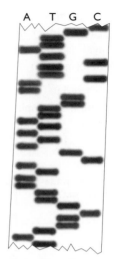

The dark bars on the gel represent fragments that ended with that particular base.

The bars near the top represent fragments that travelled further — that is, smaller fragments — than those near the bottom.

The sequence of bars therefore tells us the sequence of bases in the original DNA. Here, reading from the bottom up, the sequence is: TATGGCG ...

Interpreting a DNA sequencing electrophoresis gel

Cystic fibrosis

Cystic fibrosis (**CF**) is a genetic condition resulting from a mutation in a gene that codes for a transporter protein called **CFTR**. This protein lies in the cell surface membrane of cells in many parts of the body, including the lungs, pancreas and reproductive organs. The CFTR protein actively transports chloride ions out of cells.

There are several different mutations that result in changes in the CFTR protein. Some of them involve a change in just one base in the gene coding for the CFTR protein. The commonest one, however, involves the loss of three bases from the gene, meaning that one amino acid is missed out when the CFTR protein is being made. In all cases, the protein that is made does not work properly.

Normally, chloride ions are transported out of the cells through the CFTR protein. Water follows by osmosis. When the CFTR protein is not working, this does not happen. There is therefore less water on the outer surface of the cells than there should be. The mucus that is produced in these areas therefore does not mix with water in the usual way. The mucus is thick and sticky. As a result:

- The abnormally thick mucus collects in the lungs, interfering with gas exchange and increasing the chance of bacterial infections.
- The pancreatic duct may also become blocked with sticky mucus, interfering with digestion in the small intestine.
- Reproductive passages, such as the vas deferens, may become blocked, making a person sterile.

Gene therapy for cystic fibrosis

Gene therapy is the treatment of a genetic disease by changing the genes in a person's cells. It is only suitable for:

- diseases caused by a single gene;
- diseases caused by a recessive allele of a gene;
- serious diseases for which treatment is limited and no other cure is possible.

Although attempts have been made to treat several different diseases using gene therapy, there are still many problems to be solved before treatments become widely available and successful.

For example, attempts have been made to treat cystic fibrosis by introducing the normal CFTR gene into a person's cells. Two methods have been trialled:
- inserting the normal gene into a harmless virus and then allowing the virus to infect cells in the person's respiratory passages — the virus enters the cells and introduces the gene to them
- inserting the gene into little balls of lipid and protein, called liposomes, and spraying these as an aerosol into a person's respiratory passages

The virus and the liposomes are said to be **vectors** — they transfer the gene into the person's cells.

In each case there was some success, in that some of the cells lining the respiratory passages did take up the gene. Because the normal gene is dominant, there only needed to be one copy in a cell for it to produce normal mucus. There is no need to remove the faulty allele first, because it is recessive.

Problems with gene therapy
There were problems with the trials of gene therapy for cystic fibrosis, including:
- Only a few cells took up the normal gene, so only these cells produced normal mucus.
- It was only possible for cells in the respiratory passages to take up the normal gene, not cells in the pancreas or reproductive organ.
- Cells in the surfaces of the respiratory passages do not live for very long, so treatment would need to be repeated every few weeks.
- When using the virus as a vector, some people developed serious lung infections.

Screening for genetic conditions

Finding out the genes that a person has is called **genetic screening**.

Genetic screening can be used:
- to identify people who are carriers, that is who have a copy of a harmful recessive allele, such as the cystic fibrosis allele; a couple with cystic fibrosis in the family could therefore find out if they are both heterozygous and therefore might have a child with cystic fibrosis;
- in preimplantation genetic diagnosis, to check the genes of an embryo produced *in vitro* (that is, by fertilisation outside the body — 'test tube baby') before it is placed in the mother's uterus; this can ensure that only embryos that do not have the genes for a genetic disease are implanted;

- for prenatal testing, that is checking the genes of an embryo or fetus in the uterus; this could enable the mother to decide to have her pregnancy terminated if the baby would have a genetic disease;
- to identify people who will develop a genetic condition later in life; for example, Huntington's disease is caused by a dominant allele, but does not manifest itself until middle age; a person with this disease in the family could check if they have the gene before they decide to have children themselves;
- to identify people with alleles that put them at risk of developing other diseases; for example, a woman who has relatives with breast cancer could find out if she has the *BRAC* alleles that are known to be associated with an increased risk of breast cancer, and could decide to have a mastectomy to stop the possibility of her developing the illness.

A **genetic counsellor** will help people to interpret the results of the screening and help them to make decisions. This can be very difficult, and involve moral and ethical as well as scientific issues. For example, if a pregnant woman finds that her child will have cystic fibrosis, is this sufficient grounds to have her pregnancy terminated, or does the child have a right to life?

S Biotechnology

Biotechnology does not have a single, universal definition, but it is generally considered to mean the industrial use of microorganisms.

Extraction of metals from ores

An ore is a naturally occurring metal compound, from which the metal can be extracted. Several valuable metals, including copper, zinc and uranium, are often found in the form of sulfides. These are insoluble in water. This makes it difficult to extract the metal from the ore.

Several types of bacteria, such as *Acidothiobacillus ferrooxidans*, are able to obtain energy by changing sulfides to soluble sulfates. This is an oxidation reaction.

The sulfide ore is crushed and the bacteria are added. Water is allowed to flow through the ore, washing out the now-soluble metal sulfates the bacteria have produced. The metal can then be extracted from the metal sulfate solution.

Culturing microorganisms industrially

In industry, microorganisms can be grown in large, steel containers called **fermenters**.
- A culture of the microorganism is placed in the fermenter, together with the **nutrients** it requires for reproduction. These will normally include a carbohydrate

(as an energy source) and a nitrogen-containing substance (to allow the micro-organism to synthesise proteins and nucleotides) such as ammonia.

- If the organism is aerobic, **air** will be bubbled through the culture to provide oxygen.
- **Temperature** is usually controlled by passing cold water through a jacket surrounding the fermenter; this is because the metabolic reactions of the micro-organism are often exothermic and would cause temperature to increase.
- **pH** is controlled by using buffers.

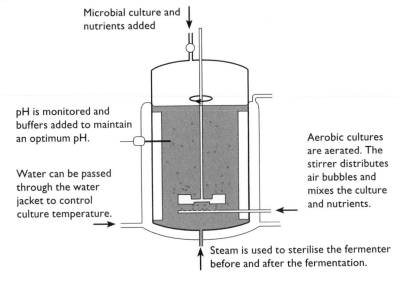

Microbial culture and nutrients added

pH is monitored and buffers added to maintain an optimum pH.

Water can be passed through the water jacket to control culture temperature.

Aerobic cultures are aerated. The stirrer distributes air bubbles and mixes the culture and nutrients.

Steam is used to sterilise the fermenter before and after the fermentation.

A fermenter

Batch culture and continuous culture

In **batch culture**, the culture is set up and the microorganism is allowed to reproduce until the nutrients have been used up. The contents of the fermenter are then harvested, and the fermenter is cleaned and sterilised before the next batch is set up.

In **continuous culture**, the culture is set up and provided with a steady supply of nutrients. The contents of the fermenter are continuously harvested. This can go on for a long time before there is a need to stop the fermentation, sterilise the fermenter and set up a new culture.

The advantages of batch culture are:

- The fermenter is totally emptied and sterilised at regular intervals, reducing the chance of contamination with other bacteria.
- If the culture does become contaminated, or if there is some other problem, then it is relatively easy to dispose of the culture and start again without too much wastage.
- After sterilisation, the fermenter can be used for a completely different process.

The advantages of continuous culture are:

- Once the culture has been set up, it can run continuously with no 'down time'.

- The continuous harvesting means there is a steady supply of the product, allowing machinery involved in its processing to be kept running continuously, and providing a steady supply of the product for sale.

Production of proteases

Protease enzymes are used in many processes and products, for example:
- in washing powders, to break down protein-based stains such as blood;
- in food processing, for example to break down proteins in meat to make it tender;
- in cheese production, where the protease rennin is used to break down casein in milk.

Bacteria such as *Bacillus stearothermophilus* can be cultured in fermenters, where they produce and secrete proteases. This is generally done by batch culture.

Production of biomass

Sometimes, the entire organism that is grown in a fermenter is required, not just a substance that it secretes. One example is **mycoprotein**.

Mycoprotein is a mass of fungal hyphae, which can be used as a food for humans or animals. The main fungus that is used for this is *Fusarium venenatum*. It is grown by continuous culture, in a process that runs for about five weeks before the fermenters are emptied and sterilised. The fermenter does not contain moving paddles, as these would break up the fungal hyphae or become entangled in them. The main nutrients supplied are glucose and ammonia.

Production of penicillin

Penicillin is an antibiotic that is produced by the fungus **Penicillium**. Penicillin is a **secondary metabolite** — that is, a substance that is produced only during certain phases of the growth of a *Penicillium* culture. It is therefore produced by batch culture.

In fact, most penicillin is produced by a variant of batch culture known as **fed batch culture**, in which nutrients are supplied continuously for a while, as in continuous culture, before the batch is harvested and a new culture set up.

How penicillin works

Penicillin is an inhibitor of a group of enzymes called glycopeptidases. These enzymes are used by bacteria to form cross-links between the peptidoglycan molecules that make up their cell walls. In the presence of penicillin, these cross-links cannot form, and the cell wall loses its strength. When the bacterium takes up water by osmosis and the cell swells, there is nothing to stop it bursting.

Penicillin does not harm viruses or human cells, as they do not have cell walls or contain peptidoglycans.

Development of resistance to penicillin

The use of antibiotics such as penicillin produces selection against bacteria that are susceptible to it.

In a population of bacteria, there will normally be some genetic variation. One or more individuals may have a gene that makes the bacterium resistant to the antibiotic. These genes are often present in plasmids, rather than in the main bacterial chromosome.

For example, some bacteria have a gene that codes for the production of β lactamase. This enzyme breaks the penicillin molecule apart.

When their environment contains penicillin, these resistant bacteria have a strong selective advantage. They are far more likely to survive and reproduce than the other members of the population. Their offspring will all contain the β lactamase gene. The next generation therefore contains a higher proportion of individuals that are resistant to penicillin.

The more often antibiotics are used, the more likely it is that populations of bacteria that are resistant to them will arise. Certain antibiotics are generally kept in reserve, to be used only when all other antibiotics have failed in the treatment of a bacterial infection.

Immobilised enzymes

Enzymes are powerful catalysts, and have many uses in industry in producing desired products from substrates.

In many production processes, the enzymes are **immobilised** by fixing them to a solid or semi-solid surface. For example, the enzyme can be trapped in beads of jelly-like calcium alginate.

> **Immobilising an enzyme in alginate**
> Make up a solution of sodium alginate.
>
> Add enzyme solution to the sodium alginate and mix thoroughly.
>
> Take up some of this mixture into a syringe, then carefully allow drops to fall into a solution of calcium chloride.
>
> The calcium chloride reacts with the sodium alginate to form little spheres of jelly-like calcium alginate. The enzyme molecules are trapped in the jelly.
>
> For further information, see **http://www-saps.plantsci.cam.ac.uk/prac_biotech.htm**

Advantages of using immobilised enzymes

- The enzyme remains fixed to the gel or other material, rather than mixing freely with the product. The product is therefore enzyme-free.
- The enzymes can be reused many times, rather than being lost with the product.

- Immobilised enzymes generally have a wider range of pH and temperature over which they can act without becoming denatured.

Dip sticks for measuring glucose concentration

Glucose concentration in a liquid, such as blood or urine, can be measured using dip sticks containing immobilised enzymes.

A small pad at one end of the dip stick contains immobilised **glucose oxidase** and **peroxidase**. It also contains potassium iodide chromogen.

When the pad is in contact with glucose, the glucose oxidase converts the glucose to gluconic acid and hydrogen peroxide.

The peroxidase then catalyses a reaction of the hydrogen peroxide with the potassium iodide chromogen. Different colours are produced according to the quantity of hydrogen peroxide formed, which in turn depends on the original concentration of glucose in the liquid.

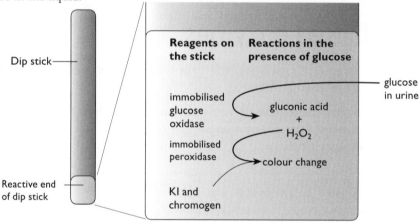

How a glucose dip stick works

Biosensors for measuring glucose concentration

A biosensor for glucose contains immobilised glucose oxidase. When in contact with glucose, the enzyme converts the glucose to gluconic acid. During this reaction, an enzyme cofactor is reduced. The reactions cause a flow of electrons, producing a tiny current that is measured and gives a digital readout.

Monoclonal antibodies

Antibodies are glycoproteins (immunoglobulins) that have specific binding sites for particular antigens. They are secreted by plasma cells.

Monoclonal antibodies are large quantities of identical antibodies produced by a clone of genetically identical plasma cells. They have many uses in medicine (see below).

Producing monoclonal antibodies

Plasma cells are not able to divide, so it is not possible to make a clone of them. A plasma cell is therefore fused with a cancer cell, to produce a **hybridoma** cell. This cell retains the ability of the plasma cell to secrete a particular antibody, and also the ability of the cancer cell to divide repeatedly to produce a clone.

- A mouse or other organism is injected with an antigen that will stimulate the production of the desired antibody.
- B lymphocytes in the mouse that recognise this antigen divide to form a clone of plasma cells able to secrete an antibody against it.
- B lymphocytes are then taken from the spleen of the mouse. These are fused with cancer cells to produce hybridoma cells.
- To identify the hybridoma cells that produce the desired antibody, individual cells are separated out. Each hybridoma cell is tested to see if it produces antibody on exposure to the antigen. Any that do are cultured so that they produce a clone of cells all secreting the desired antibody.

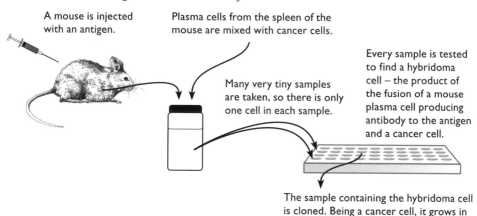

A mouse is injected with an antigen.

Plasma cells from the spleen of the mouse are mixed with cancer cells.

Many very tiny samples are taken, so there is only one cell in each sample.

Every sample is tested to find a hybridoma cell – the product of the fusion of a mouse plasma cell producing antibody to the antigen and a cancer cell.

The sample containing the hybridoma cell is cloned. Being a cancer cell, it grows in unlimited quantities in the laboratory.

How monoclonal antibodies are produced

Uses of monoclonal antibodies

Monoclonal antibodies can be used in the diagnosis of disease, and in pregnancy testing.

Their big advantage is that the tests can be carried out quickly, often producing results in minutes or hours.

Diagnosis

Monoclonal antibodies can be used in an ELISA test to diagnose an infectious disease, by detecting the presence of particular antigens in the blood. ELISA stands for enzyme-linked immunosorbent assay.

1 The monoclonal antibody is immobilised on the surface of a small container, such as a small glass well. The liquid to be tested (for example, blood plasma) is added to the well. If the antigens are present, they will bind to the antibodies.

2 The contents of the well are then rinsed out. The antigens stay tightly bound to the antibodies.

3 Now another solution containing the same monoclonal antibodies is added to the well. These antibodies also have a reporter enzyme combined with them. They bind with the antigens already attached to the antibodies in the well. The well is again rinsed out, so the enzymes will all be washed away unless they have bound with the antigens.

4 The substrate of the enzyme is then added. If the enzyme is present — which is only the case if the antigen being investigated was present — then the substrate is changed to a coloured substance by the enzyme.

The colour change therefore indicates the presence of the antigen in the fluid being tested.

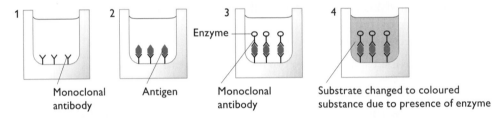

How an ELISA test works

Pregnancy testing

The urine of a pregnant woman contains the hormone human chorionic gonado-trophin, hCG. Monoclonal antibodies can be used to detect its presence in urine.

There are many different types of pregnancy testing kits. One type uses a plastic strip containing three bands:
- The R band contains immobilised monoclonal antibodies that can bind with hCG. These antibodies have been combined with an enzyme.
- The T band contains more antibodies that can bind with hCG, and also coloured substrates for the enzymes in the R band.
- The C band is used to check that the strip is working.

When the end of the strip is dipped into urine, the urine moves up the strip by capillary action. If it contains hCG, this binds with the monoclonal antibodies in the R band. The complexes of hCG, antibodies and enzymes break free from the strip, and continue moving up it as the urine seeps upwards.

When they reach the T band, the enzymes attached to the antibodies cause the substrate to react, producing a coloured product. This produces a coloured stripe on the test strip.

Treating disease

Monoclonal antibodies can be produced that will bind with particular proteins on the surface of body cells. For example, the monoclonal antibody rituximab binds with a protein called CD20, which is found only on B lymphocytes. This can be useful in the treatment of a type of cancer called non-Hodgkin lymphoma, in which B cells divide uncontrollably. When rituximab binds to the B cells, it makes them 'visible' to the immune system, which destroys them. New B cells are made in the bone marrow, and these replacement cells may not be cancerous.

T Crop plants

Crop plants are plants that are grown by humans for food and other resources. In this section, we will consider four **cereal crops** — **maize** (corn), **sorghum**, **rice** and **wheat**. A cereal crop is a grass-like plant that is grown for its seeds.

Maize is grown in tropical and temperate climates. It is able to grow well in high temperatures, but varieties have also been bred that can grow in cooler climates.

Sorghum is able to grow in hotter, drier conditions than other cereal crops.

Rice is mostly grown in tropical and sub-tropical climates, as it generally requires temperatures of at least 20 °C. Rice is adapted to grow in wet conditions.

Wheat is grown in temperate climates. It is able to survive temperatures well below freezing.

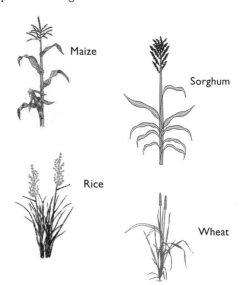

Flower and fruit structure

Flowers are the organs in plants in which sexual reproduction takes place.

The male gametes are contained in pollen grains, which are made in anthers attached to long filaments. The female gametes are contained in ovules, inside ovaries.

At the top of the ovary is a long style, with a stigma at the top. During pollination, pollen grains are deposited on the stigma.

Wind pollination

All cereal crops have flowers that are pollinated by the wind. The structure of a maize plant and its flowers shows features characteristic of wind-pollinated plants. Maize plants have separate male flowers and female flowers, both on the same plant.

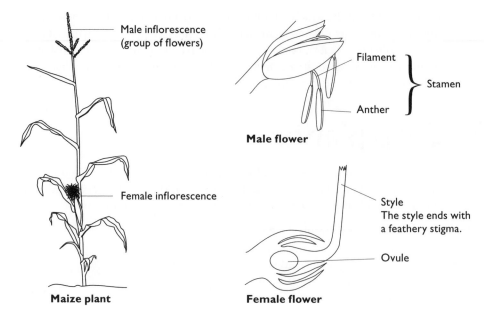

Male inflorescence (group of flowers)

Filament

Stamen

Anther

Male flower

Female inflorescence

Style
The style ends with a feathery stigma.

Ovule

Maize plant

Female flower

A maize flower — an example of a wind-pollinated flower

- The flowers are on long stalks, held high so that the wind can catch them.
- The male flowers are above the female flowers, enabling pollen to fall onto the stigmas below.
- The anthers dangle out of the flower on long, flexible filaments, making it easy for the pollen to be blown away.
- The pollen is made in large quantities and the grains are small and smooth, so that they can be easily carried by the wind.
- The stigmas are large and feathery, providing a large surface area to catch the wind-blown pollen.
- There are no petals, as there is no need to attract insects, and petals would shield the anthers and stigmas from the wind.

Self-pollination and cross-pollination

Self-pollination occurs when pollen from one flower lands on the stigma of the same flower, or of another flower on the same plant.

Cross-pollination occurs when pollen from one flower lands on the stigma of a flower on a different plant of the same species.

After pollination, a tube grows from the pollen grain down through the style. The haploid male gamete travels down the tube into the ovule, where it fuses with the haploid female gamete to form a diploid zygote.

If the male gamete came from the same plant as the female gamete (that is, if self-pollination occurred), then all of the alleles in the zygote will come from its single parent. There is only a limited possibility of variation. It is also possible that, if the parent was heterozygous at any gene locus, then the zygote will receive two reces-sive alleles for that gene. This could produce undesirable features in the offspring.

If the male gamete came from a different plant (that is, if cross-pollination occurred), then the zygote will contain alleles from both parents. If the parents were not geneti-cally identical, then there is likely to be much greater genetic variation amongst the offspring than would occur with self-pollination.

Fruit formation

After pollination and fertilisation, the fertilised ovule develops into a **seed**. The zygote develops into an **embryo plant**, inside the seed.

Other tissues in the ovary develop into the **endosperm** tissue. This contains food stores, mostly starch, which will be used by the embryo when the seed germinates.

Around the edge of the endosperm is the aleurone layer. This contains enzymes that are activated when the seed germinates. The enzymes break down the starch in the endosperm (page 172).

The ovary develops into a **fruit** (the maize grain), with the seeds inside it. In maize, each ovary contained a single ovule, so each fruit contains a single seed. The outer layer of a seed is the **testa**, and the fruit wall is the **pericarp**. In a maize grain, the testa and pericarp are fused together.

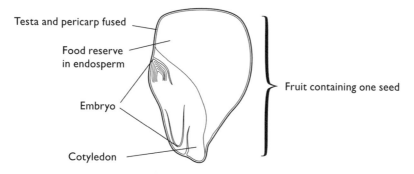

Structure of a maize fruit

Cereal grains in the human diet

All cereal grains have a structure similar to that of the maize fruit. Cereal grains are **staple foods** in many parts of the world. They contain:
- large amounts of carbohydrate, mostly in the form of starch (in the endosperm)
- protein, mostly in the aleurone layer and the embryo
- only small amounts of lipids
- vitamin B
- calcium
- fibre (cellulose and lignin)

Cereal grains are dry, and can therefore be stored for long periods.

Photosynthesis in maize and sorghum

Maize and sorghum are **C4 plants**. This means that, instead of first making a 3-carbon compound during the Calvin cycle (page 151), they produce a 4-carbon compound. This is an adaptation to growing in environments where the temperature and light intensity are high.

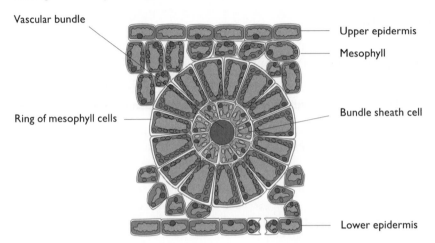

Section through a leaf of a C4 plant

At high temperatures and high light intensities, the enzyme rubisco tends to cata-lyse the combination of RuBP with oxygen rather than with carbon dioxide. This is wasteful, and reduces the rate of photosynthesis. It is sometimes called 'photorespi-ration', because it uses oxygen. The leaves of C4 plants have structural adaptations that prevent photorespiration taking place.

In a C4 plant, rubisco and RuBP are kept away from the air spaces inside the leaf, so they do not come into contact with oxygen. The rubisco and RuBP are inside the chloroplasts of the **bundle sheath cells**. They are separated from the air spaces by a ring of **mesophyll cells**. These also contain chloroplasts, where the light-dependent reactions of photosynthesis take place.

Carbon fixation happens like this:
- In the mesophyll cells, carbon dioxide combines with PEP to form a 4-carbon compound.
- The 4-carbon compound moves into the bundle sheath cells.
- The 4-carbon compound breaks down and releases carbon dioxide.
- The rubisco in the bundle sheath cells catalyses the reaction of the carbon dioxide with the RuBP.
- The Calvin cycle then proceeds as normal, inside the bundle sheath cells.

The enzymes involved in photosynthesis in maize and sorghum have **higher optimum temperatures** than in plants that are not adapted for growth in hot climates.

Adaptations of sorghum for arid environments

Both maize and sorghum are widely grown in arid climates, for example in regions of sub-Saharan Africa where there may be little rain during the growing season. Sorghum can be grown even where it is too dry to grow maize.

Adaptations of sorghum for growth in dry conditions

Feature	How it helps the plant to survive in dry conditions
Sorghum plants have a relatively small leaf area.	This reduces the area from which water can evaporate in transpiration, therefore reducing the rate of water loss.
Sorghum leaves and internodes are covered with a layer of wax.	This is impermeable to water and therefore decreases water loss.
Sorghum leaves have a row of motor cells along the midrib that allow the leaves to roll up when the cells are short of water.	This decreases the surface area of the leaves in contact with air, and therefore reduces the rate of loss of water vapour from the leaves to the air. Moist air is trapped inside the rolled leaf.
Sorghum leaves have relatively few stomata, and these are sunken below the leaf surface.	Moist air, with a high water potential, is trapped around the stomata. This reduces the water potential gradient between the air spaces in the leaf and the outside, reducing the rate of loss of water vapour from the leaf.
The root system is extensive and finely branched.	The roots are able to absorb water even when there is very little water in the soil.
The plant can close its stomata and become dormant for long periods.	The plant is able to survive during a prolonged drought, resuming growth when conditions improve.

Adaptations of rice for wet conditions

Rice is often grown in fields that are flooded with water during the growing season.

Adaptations of rice for growth with roots submerged in water

Feature	How it helps the plant to survive when roots are submerged
Cells are tolerant of high concentrations of ethanol.	When roots are submerged in water, less oxygen is available than when the soil contains air spaces. Cells therefore respire anaerobically, producing ethanol.
Stems have tissues called aerenchyma, containing large air spaces.	Aerenchyma allows oxygen from the air to diffuse down to the roots.
Some types of rice are able to grow elongated stems to keep their leaves above the water as its level rises.	The leaves remain exposed to the air, which facilitates gas exchange for photosynthesis and respiration.

Crop improvement

New varieties of crops are produced by both conventional breeding techniques (selective breeding) and genetic modification.

Conventional breeding techniques

Polyploidy in wheat

The original ancestors of wheat are grasses that grow wild in the Middle East. There is some doubt about exactly how modern wheat developed. One theory is as follows.

- About 0.5 million years ago, *Aegilops speltoides* (goat grass) formed a hybrid with *Triticum urartu* (einkorn).
- Hybrids between species are usually sterile, because the two sets of chromosomes in a diploid cell do not match and therefore cannot pair up and undergo meiosis to form gametes. However, chromosome doubling occurred in the hybrid to produce a plant with four sets of chromosomes in its cells — two sets from the *A. speltoides* parent and two sets from the *T. urartu* parent. The plant was therefore tetraploid (4n) and could undergo meiosis.
- This tetraploid plant is *Triticum turgidum*, emmer wheat.
- About 9000 years ago, emmer wheat formed a hybrid with another wild goat grass, *Aegilops tauschii*. Chromosome doubling occurred again, resulting in a hexaploid (6n) plant, *Triticum aestivum*. This is modern wheat. It has two sets of chromosomes from each of its three parents — *A. speltoides*, *T. urartu* and *A. tauschii*.

In wheat, polyploids (plants with more than two complete sets of chromosomes) grow more vigorously and have higher yields of grain than diploid plants.

Inbreeding and hybridisation in maize

Farmers require crop plant varieties in which all the individuals are genetically identical. This means that, if the seed is sown at the same time and in the same conditions, then the plants will all grow uniformly. They will ripen at the same time and grow to the same height, which makes harvesting easier. The grain will all have similar characteristics, which makes marketing easier.

Genetic uniformity is usually achieved through **inbreeding** (breeding a plant with itself, or with other plants with the same genotype) for many generations. However, in maize, inbreeding results in weak plants with low yields. This is called **inbreeding depression**.

Maize breeding therefore involves producing **hybrids** between two inbred lines. Like most selective breeding of cereal crops, it is done by commercial organisations, not by farmers themselves.

- Inbred lines (genetically uniform populations) of maize with desirable characteristics are identified, and crossed with other inbred lines with different sets of desirable characteristics.
- The resulting hybrids are assessed for features such as yield, ability to grow in dry conditions, and resistance to insect pests. The best of these hybrids are then chosen for commercial production.

- Large quantities of the two inbred lines from which the hybrid was bred are grown, as it is from these that all the seed to be sold will be produced.
- Each year, the two inbred lines are bred together, and the seed collected from them to sell as hybrid seed.

This method of breeding avoids problems of inbreeding depression. The hybrid plants are genetically uniform (although they will be heterozygous for several genes) because they all have the same two inbred parents.

Using genetic modification in crop breeding

Herbicide-resistant oil seed rape

Oil seed rape, *Brassica napus*, is a relative of cabbages that is widely grown in temperate countries for its oil-rich seeds.

Genes that confer resistance to herbicides (weedkiller) containing glyphosate and glufosinate have been inserted into oil seed rape cells. Oil seed rape plants containing this gene do not die when this herbicide is sprayed on them. This allows farmers to control weeds in their crops by spraying the field with glufosinate, which kills all the plants except the oil seed rape.

There are some potential detrimental effects. Genes from the GM (genetically modified) oil seed rape plants might be able to spread into wild *Brassica* plants growing nearby, and make them resistant to the herbicides. Oil seed rape is able to hybridise with wild relatives, which are often found growing close to rape fields. However, after several years in which the GM oil seed rape has been grown on a very large scale in Canada, and in trials in Europe, only a very few such hybrids have been found, and when tested were found to have produced seeds that could not germinate. Hybrids between different species are usually infertile, which makes it unlikely that they would be able to produce populations of resistant plants. However, the possibility of the resistance genes spreading into wild populations of plants does remain.

Insect-resistant maize and cotton

Yields of maize can be greatly reduced by an insect larva called the corn borer. Cotton yields are reduced by the cotton boll worm, which is also an insect larva.

Pesticides sprayed onto the crops can kill these insects. However, the pesticides can also harm other, beneficial insects. The insects also evolve resistance to the pesticides.

Genes that code for the production of a protein derived from the bacterium *Bacillus thuringiensis*, called Bt toxin, have been inserted into maize and cotton plants. The plants therefore produce the protein, which is converted into the toxin once inside the gut of insects that have eaten the leaves. This means that the toxin kills insects that feed on the plants, but not other insects.

Benefits include:
- less loss of the crop to insect pests, so greater yields are obtained which can keep costs down;
- the toxin harms only insects that eat the plant, not other insects as happens when insecticides are sprayed on crops;

- it is less likely that insect pests will evolve resistance to the Bt toxin than to pesticides. However, there are signs that resistance can develop in some pest species. This can be counteracted by using slightly different forms of the Bt toxin, or a combination of two different Bt toxins, in GM crops.

Possible detrimental effects include:
- the seed of Bt corn and cotton costs more for farmers to buy than non-GM seed, making it difficult for farmers in developing countries to afford it, and therefore making it difficult for them to compete with farmers in richer countries;
- it is possible that some non-pest insects might be harmed by the Bt toxin. However, overall it is found that there are more non-pest insects present in fields where Bt corn or Bt cotton are grown than in fields where non-Bt crops are grown and sprayed with pesticides.

Vitamin A enhanced rice

Rice is a staple food in many parts of the world. It does not contain much vitamin A, and children whose diet consists largely of rice may suffer deficiency diseases such as night blindness, or lack of resistance to infectious diseases. A new variety of rice, called Golden Rice™, has been genetically modified to produce large quantities of β carotene in its grains, which human cells can convert to vitamin A. The first version of this rice used genes taken from daffodils and a bacterium *Erwinia uredovora*, but the quantity of β carotene produced was not very large. Since then, new varieties containing genes from maize and rice itself have been produced, which contain up to 31 µg of β carotene per 100 g of rice. The rice has to undergo further tests, for example to check its effects on human health, before it will become widely available. It will also be necessary to incorporate the genes into different varieties of rice that are suitable for growing in different parts of the world.

Potential benefits include:
- children whose diet consists largely of rice would be able to get more than half of their daily requirement for vitamin A by eating 200 g of this rice per day;
- the researchers who developed this GM rice will donate it free of charge for use in developing countries.

Possible detrimental effects include:
- the existence of this GM rice could possibly lessen efforts to tackle the root causes of poverty and poor diet in some parts of the world;
- some people argue that the changes made to the genes in the rice could have harmful effects on people who eat it; however there is no evidence for this.

U Aspects of human reproduction

Humans reproduce sexually, producing male and female haploid gametes whose nuclei fuse at fertilisation to form a diploid zygote. The zygote develops into an embryo and then a fetus in its mother's uterus, attached to the uterus wall by a placenta, through which it obtains nutrients and oxygen and gets rid of urea and carbon dioxide.

Gametogenesis

Gametogenesis is the production of haploid gametes from diploid somatic (body) cells. Spermatogenesis takes place in the testes, and oogenesis in the ovaries.

Spermatogenesis

Diploid spermatogonia at the edge of the seminiferous tubule undergo mitosis and then meiosis to produce haploid spermatids. As meiosis proceeds, the cells move towards the centre of the tubule. The whole process takes about 64 days.

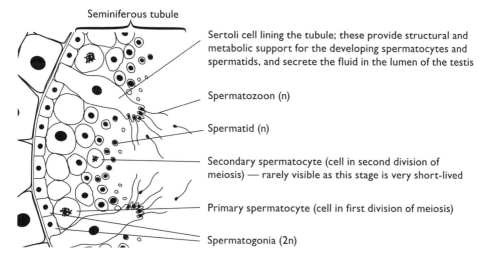

Histology of the testis

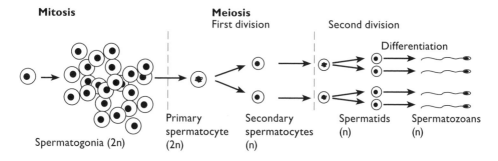

Sequence of spermatogenesis

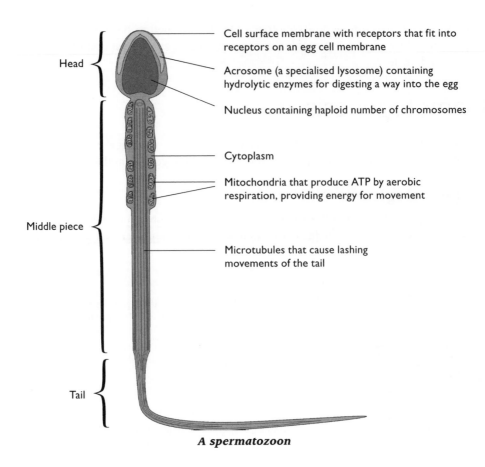

Head {

Cell surface membrane with receptors that fit into receptors on an egg cell membrane

Acrosome (a specialised lysosome) containing hydrolytic enzymes for digesting a way into the egg

Nucleus containing haploid number of chromosomes

Cytoplasm

Mitochondria that produce ATP by aerobic respiration, providing energy for movement

Middle piece {

Microtubules that cause lashing movements of the tail

Tail {

A spermatozoon

Oogenesis

Diploid oogonia divide by mitosis and then meiosis to produce haploid secondary oocytes. At birth, the oogonia have already become primary oocytes, in the first division of meiosis. The primary oocytes are inside primordial follicles, in which the oocyte is surrounded by a layer of granulosa cells. The primordial follicle remains in this state for many years.

At puberty, some of the primordial follicles develop into primary follicles, in which the primary oocyte grows larger and develops a coat called the zona pellucida. The primary follicle then develops into a secondary follicle, containing extra layers of granulosa cells, which are surrounded by a theca.

From puberty onwards, some of the primary follicles develop into ovarian follicles. These contain a large fluid-filled cavity, the antrum, between the granulosa cells. Just before ovulation, the primary oocyte completes the first division of meiosis to produce a secondary oocyte and a tiny polar body. At ovulation, the secondary oocyte is in metaphase of the second division of meiosis. After ovulation, the remains of the ovarian follicle develop into a corpus luteum.

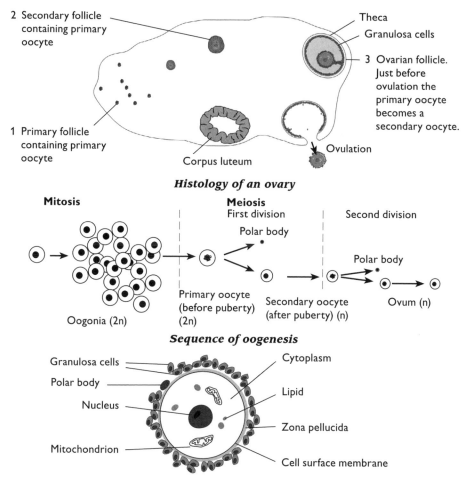

Histology of an ovary

Sequence of oogenesis

A secondary oocyte and polar body at ovulation

The menstrual cycle

The menstrual cycle is a repeating process of change in the **ovaries**, **oviducts** and **uterus**, which takes place approximately every 28 days from puberty to menopause. It is controlled by four hormones:

- two **steroid hormones** secreted by the ovaries — **oestrogen** and **progesterone**;
- two **gonadotropic peptide hormones** secreted by the **anterior pituitary gland** — **FSH** (follicle stimulating hormone) and **LH** (luteinising hormone).

First half of the cycle

During the first half of each cycle, the dominant gonadotropin is FSH. This stimulates secondary follicles in the ovary to grow in size and number. The granulosa cells of the secondary follicles secrete oestrogen, which is the dominant steroid hormone in this stage of the cycle. This affects:

- the oviduct, causing the development of more, larger and more active cilia on the cells lining its walls. It also increases the secretion of more glycoprotein; these changes prepare the oviduct for the arrival of an oocyte, which will be moved along the oviduct by the cilia;
- the endometrium (lining of the uterus), causing the cells to divide to form a thicker layer, ready to receive an embryo if the egg is fertilised.

Mid-point of the cycle

At day 13 or 14, a surge in LH causes the primary oocyte in a single ovarian follicle to complete meiosis I and continue to metaphase of meiosis II. It also causes the follicle to shed the secondary oocyte into the oviduct. This is ovulation.

Second half of the cycle

The LH surge also causes the granulosa cells lining the secondary follicles to change to luteal cells, and switch to secreting largely progesterone and less oestrogen.

Progesterone becomes the dominant steroid hormone from 5 days after ovulation (which is when a fertilised embryo would enter the uterus). Progesterone causes the cells in the endometrium to differentiate to form a thick, vascularised layer. If fertilisation does not occur, the corpus luteum begins to degenerate from day 23, so the secretion of progesterone also decreases. This causes the breakdown of the endometrium, leading to the start of menstruation on day 0 of the start of the next cycle.

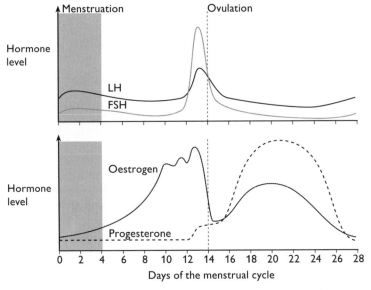

Changes in hormones during the menstrual cycle

Negative feedback in the menstrual cycle

High concentrations of oestrogen inhibit the secretion of FSH and LH by the anterior pituitary gland. This happens during the first half of the menstrual cycle, causing the levels of FSH and LH to fall. However, when oestrogen levels are very high, a surge of LH secretion occurs, which brings about ovulation.

Towards the end of the cycle, as oestrogen and progesterone levels fall, the inhibition of FSH and LH secretion is lifted, so the concentration of these two hormones begins to increase, leading to the start of a new cycle.

Contraception

The contraceptive pill

Most contraceptive pills contain synthetic hormones similar to progesterone and oestrogen. These prevent the secretion of LH and FSH from the anterior pituitary gland. There is therefore no development of secondary follicles in the ovary (which is caused by FSH) and no ovulation (which is caused by a surge of LH).

Potential advantages of the use of contraception

- Family size can be kept to the level desired by the parents, which could potentially reduce poverty and make it more possible for each child to have a better standard of living, including better nutrition, health care and education.
- Partners who do not wish to have children for reasons such as the risk of inheriting a genetic disorder can still have an active sex life.
- Use of the contraceptive pill can enable women to determine their own fertility, so they can avoid having an unwanted child even if their partner does not take this responsibility.
- Use of contraception avoids unwanted pregnancies and therefore reduces the number of abortions; abortion is stressful for a woman (and her partner) and is held to be morally undesirable by many people.

Potential disadvantages of the use of contraception

- With no fear of becoming pregnant, a woman may be more prepared to have sexual intercourse with more partners; there is evidence that the widespread availability of contraception has increased promiscuity among young people.
- Sexual intercourse with more partners increases the risk of the spread of diseases such as HIV/AIDS, and can also increase the risk of marriage breakdown or stress.
- A man may feel that he can have intercourse with his female partner whenever he wishes, even if she does not want this.

In-vitro fertilisation

'In-vitro' means 'in glass', and refers to the fact that fertilisation occurs in glassware (for example a Petri dish) in the laboratory, rather than in a woman's oviduct.

- The woman is given hormones to induce the development of secondary follicles and ovarian follicles in her ovaries. These may include LH and FSH. The dosage is large enough to cause several follicles to develop simultaneously.
- Once the follicles have developed, the woman will be given the hormone human chorionic gonatropin (hCG). This will stimulate the formation of a corpus luteum, which will secrete progesterone. This is necessary to prepare the woman's oviduct and endometrium to receive an embryo.
- Oocytes will then be harvested from her ovaries, under general anaesthetic.
- On the same day, semen is collected from the woman's partner. The sperm are washed and placed in a fluid that contains nutrients that will enable the sperm to become ready to fertilise an egg (capacitation).
- Each egg is placed in a separate dish, and sperm are added. Any embryos produced are ready to be transferred to the woman's uterus after three to five days.

Ethical implications of IVF

For many women and their partners, IVF makes it possible for them to have a family when otherwise they could not. However, there are some potential negative aspects that need to be considered. Points of view of different individuals can be widely different, sometimes because of religious or cultural viewpoints, but sometimes because of a person's own moral viewpoint. Some of these issues are outlined below.

- Generally, two five-day old embryos are transferred. However, some doctors prefer to transfer more embryos. This can lead to multiple embryos developing in the uterus, which are less likely to be healthy than a single one.
- It is possible for a woman to have children using sperm from a man who is not her partner. Some people believe this is not ethically acceptable. There are also difficulties if the man who donated the sperm does not want any resulting children to know that he is the father.
- It is possible for a woman who is long past child-bearing age and who has reached the menopause to have children in this way. This could mean that a child is born to a mother who will not have a sufficiently long active life to care for him or her.
- While many people think it is important that a woman should have the right to bear a child, others feel that there are enough children in the world already and that a person who is biologically unable to have one should not be treated, especially if the cost is borne by tax-payers or if her treatment prevents the treatment of someone else with a life-threatening condition.
- Some of the embryos resulting from IVF will not be implanted. They could be allowed to die, or they could be used for research. Some people are concerned that we do not have the right to dispose of embryos in this way.
- It is possible for semen from a partner to be collected and frozen. A woman could then have his child by IVF after the partnership has broken up, or after he has died, without his permission.
- The embryos produced by IVF are usually tested to ensure they are viable (able to survive) before implanting them. This may include checking for the presence of undesirable alleles such as those that can cause genetic conditions like cystic fibrosis. It is also possible to choose an embryo of a particular sex. There are wide variations in people's views of the ethics of these choices.

A2
Experimental
Skills &
Investigations

A2 Experimental skills and investigations

More than one quarter of the marks in your A2 examinations are for practical skills. These are assessed on Paper 5. This is **not** a practical examination. It is a written paper. However, you will need to do plenty of practical work in a laboratory throughout your A2 biology course in order to develop the practical skills that are assessed in Paper 5.

There is a total of 30 marks available on this Paper. Although the questions are different on each Paper 5, the number of marks assigned to each skill is always the same. This is shown in the table below.

Skill	Total marks	Breakdown of marks	
Planning	15 marks	Defining the problem	5 marks
		Methods	10 marks
Analysis, conclusions and evaluation	15 marks	Dealing with data	8 marks
		Evaluation	4 marks
		Conclusion	3 marks

The syllabus explains each of these skills in detail, and it is important that you read the appropriate pages in the syllabus so that you know what each skill is, and what you will be tested on.

The next few pages explain what you can do to make sure you get as many marks as possible for each of these skills. They have been arranged by the kind of task you will be asked to do.

Some of the skills are the same as for AS. However, most of them are now a little more demanding. For example, in dealing with data, you are not only expected to be able to draw results charts and graphs, but also to use statistics to determine what your results show.

How to get high marks in Paper 5

Planning

At least one of the questions on Paper 5 will require you to plan an investigation.

The question will describe a particular situation to you, and ask you to design an experimental investigation to test a particular hypothesis or prediction. The planning question will usually involve investigating the effect of one factor (the independent variable) on another (the dependent variable).

Sometimes, the question will tell you the apparatus you should use. In other cases, you will need to decide on the apparatus. If you have done all of the practicals listed in the syllabus during your course, then you will have had direct experience of a wide range of apparatus and techniques, which will make this much easier for you.

Stating a hypothesis

You may be asked to state a hypothesis. For example, you might be asked to plan an experiment to find out how light intensity affects the rate of photosynthesis. Your hypothesis could be:

> Increasing light intensity will increase the rate of photosynthesis.

Notice that your hypothesis should include:
- reference to the independent variable (light intensity)
- reference to the dependent variable (rate of photosynthesis)
- a clear statement of how a particular change in the independent variable (an increase) will affect the dependent variable (an increase).

You could be asked to express your hypothesis in the form of a sketch graph. Here, you could draw a graph like this:

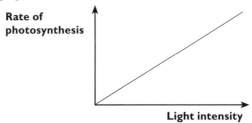

Your hypothesis should be:
- **quantifiable** — this means that you should be able to *measure* changes in the independent and dependent variables and obtain numerical results
- **testable** — this means that you should be able to do an experiment that will test whether or not your hypothesis could be correct
- **falsifiable** — this means that you should be able to do an experiment whose results could show if your hypothesis is not correct.

Notice that it is not possible to actually *prove* that a hypothesis is correct. For example, if your experiment involved putting a piece of pondweed in a tube and measuring the rate at which oxygen is given off with a lamp at different distances from the tube, you might find that the rate of photosynthesis does indeed increase with light intensity. You could say that your results 'support' your hypothesis. However, you cannot say that they 'prove' it, because you would need to do more experiments and obtain a lot more data before you could be sure this relationship is *always* true.

However, you can *disprove* a hypothesis. If you kept on increasing the light intensity in this situation, you would probably find that the rate of photosynthesis eventually levels off, as some other factor begins to limit the rate. This would disprove your hypothesis.

Variables

You will usually need to decide what are the *independent* and *dependent variables*. You will already have thought about this for AS, and it is described on page 100.

You will also need to identify important *variables that should be controlled* or standardised. This is described on page 104.

Range and intervals

You will need to specify a suitable range and intervals of the independent variable that you would use. This is described on pages 100 to 102.

Apparatus

Sometimes you will be told exactly what apparatus you should use. If so, then it is very important that your experiment does use this apparatus. If you describe a different kind of apparatus, you will not get many marks.

For example, the question may include a diagram of a plant shoot in a test-tube containing dye solution, with a layer of oil on the liquid surface. The question may ask how you could use this apparatus to find the rate of movement of water up the stem. You will probably never have used this apparatus for this purpose, but you should be able to work out a way of using it for this investigation. If you describe the use of a different piece of apparatus — for example, a potometer — then you will not get many marks, because you have not answered the question.

If you do plenty of practical work during your biology course, then you will become familiar with a wide range of apparatus. Check all the learning outcomes in the syllabus that begin with [PA], as this indicates that practical work may be involved. There is also a list of basic apparatus that you should be able to use towards the end of the syllabus, under the heading 'Laboratory equipment'.

Method

You will need to describe exactly what you will do in your experiment.

Changing and measuring the independent variable

You should be able to describe *how* you will change the independent variable and *how* you will measure it. If the independent variable is the concentration of a solution, then you should be able to describe how to make up a solution of a particular concentration.

To make up a solution of $1\,mol\,dm^{-3}$, you would place 1 mole of the solute in a $1\,dm^3$ flask, and then add a small volume of solvent (usually water). Mix until fully dissolved, then add solvent to make up to exactly $1\,dm^3$. (1 mole is the relative molecular mass of a substance in grams.)

To make up a 1% solution, you would place 1 g of the solute in a container with a small volume of solvent. Mix until fully dissolved, then make up to $100\,cm^3$ with solvent.

You should be able to use a stock solution to make serial dilutions. This is described on pages 102 to 103. It is worth remembering that a small test-tube holds about $15\,cm^3$, and a large one (sometimes called a boiling tube) up to $30\,cm^3$.

Keeping key variables constant

You will be expected to describe *how* you will control a particular variable. For example, if you need to control temperature, then you should explain that you would use a water-bath (either thermostatically controlled, or a beaker of water over a Bunsen burner) in which you take the temperature using a thermometer.

Measuring the dependent variable

You should describe *how* you will measure the dependent variable, and *when* you will measure it.

Sequence of steps

Your method should fully describe the sequence of the steps you will carry out. This will include setting up the apparatus and all the other points described above relating to changing, measuring and controlling variables. You may like to set this out as a series of numbered points.

Risk assessment

You should be able to identify any ways in which your experiment might cause a risk of injury or accident, and to suggest safety precautions related to these risks. Sometimes these are obvious — for example, if you are using concentrated acid, then there is a risk that this might come into contact with clothing, skin or eyes, and therefore protective clothing and eye protection should be worn. In the experiment investigating the effect of light intensity on rate of photosynthesis, care should be taken not to get water on the electrical light source, as this could cause a current surge or give a person an electric shock.

Sometimes, an experiment might be so simple and safe that there are no genuine risks. If so, then you should say so. Do not invent risks if there are none!

Reliability

As described on page 116, it is a good idea to do at least **three repeats** for each value of the independent variable. A mean can then be calculated, which is more likely to give a true value than any one of the individual values.

Recording and displaying data, and drawing conclusions

This is the same as for AS. It is described on pages 107 to 114.

> **Tips** In the exam:
> - Check exactly what the question is asking you do. You may be expected to state your hypothesis both in words and in the form of a predicted graph.
> - Even if you are not asked to write them down, make sure that you have identified the independent and dependent variables.
> - State clearly which variables you would control, and be prepared to state exactly how you would do this.
> - Take care that your plan uses the apparatus that the question asks you to use, even if you think this is not the best apparatus, or if it is not apparatus that you have used before.

- Always think about risk. Identify any genuine risks and explain how you would minimise them. Do not invent risks if there are none.
- It is almost always a good idea to suggest doing at least three repeats.

Dealing with data

Tables and graphs

You may be asked to draw tables or graphs to record and display data. These are described on pages 107 to 113.

Calculations to summarise or describe data

You will be asked to carry out some type of calculation, using data that you have been given.

Mean, median and mode

These are best explained using an example. Let's say that you had measured the lengths of 20 leaves. Your results, recorded in mm, were:

33.5, 56.5, 62.0, 75.0, 36.0, 54.5, 43.5, 41.5, 54.0, 53.0
53.5, 39.0, 72.5, 66.5, 58.5, 42.5, 41.5, 49.0, 69.0, 38.5

(Notice that all the measurements are to three significant figures, or one decimal place, and that all values have been measured to the nearest 0.5 mm.)

To calculate the **mean** value, add up all the values and divide by the total number of measurements.

$$\text{mean value} = \frac{\text{sum of individual values}}{\text{number of values}} = \frac{1040}{20} = 52.0$$

Now let's say you measured the lengths of 80 more leaves, and obtained the results displayed in this histogram.

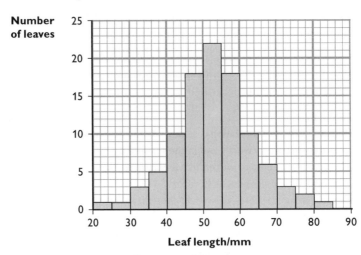

A frequency histogram

The **mode**, or **modal class**, is the most common value in the set of observations. In this instance, this is 50–54 mm. The **median** is the middle value of all the values. In this case, the median value is 52.5 mm.

If the frequency histogram of your data is a perfect bell-curve, we say that the data show a **normal distribution**. For a perfect normal distribution, the mean, mode and median are all the same.

Percentage increase or decrease
You can remind yourself how to calculate these on page 108.

Range and inter-quartile range
You have already met the idea of 'range', when thinking about the range of values you would use for an independent variable in an investigation. The **range** is the spread between the highest and lowest values in your data. For the 100 leaves in the histogram above, the range is 65 mm.

The **inter-quartile range** describes the range between the values that are one quarter and three quarters along the complete range of your data. For the leaf data, the inter-quartile range is from 36.25 to 68.75. These values tell you how much the data are spread. If they are all very close to each other, then the range and the inter-quartile range will be small.

This can be useful if you are looking at a set of repeat results in an experiment. Ideally, you would hope that all the results are very close to one another. If they are not, then this indicates a lack of reliability — there is a lot of variation in your results for a particular value of your independent variable and therefore you cannot have a great deal of faith that your mean is the 'true' value.

Statistics
Statistical methods are designed to help us to decide the level of confidence we can have in what our results seem to be telling us. This is often a problem in biology, where we are generally dealing with variable organisms and may not be able to fully control all the control variables we would like. This means that we do not get a set of results that precisely matches what we have predicted from our hypothesis, or where we are not really sure whether the set of results we have obtained for one experiment is genuinely different from the set of results we have obtained for another experiment.

In most statistical tests, you will need to calculate some basic values along the way. These include **standard deviation** and **standard error**.

Calculating standard deviation
This is used with data that show an approximately normal distribution. It is a measure of how much the data vary around the mean.

The larger the standard deviation, the wider the variation.

This is how you would calculate the standard variation for the leaf lengths shown on page 230. These were:

33.5, 56.5, 62.0, 75.0, 36.0, 54.5, 43.5, 41.5, 54.0, 53.0
53.5, 39.0, 72.5, 66.5, 58.5, 42.5, 41.5, 49.0, 69.0, 38.5

1 Calculate the mean. This is 52.0.
2 For each measurement, calculate how much it differs from the mean, and then square each of these values. It is easiest to do this if you set the results out in a table. In the table:

x represents an individual value
$\overline{X}$ represents the mean
Σ represents 'sum of'

x	$(x - \overline{X})$	$(x - \overline{X})^2$
33.5	−18.5	342.25
56.5	4.5	20.25
62.0	10.0	100.00
75.0	23.0	529.00
36.0	−16.	256.00
54.5	2.5	6.25
43.5	−8.5	72.25
41.5	−10.5	110.25
54.0	2.0	4.00
53.0	1.0	1.00
53.5	1.5	2.25
39.0	−13.0	169.00
72.5	20.5	420.25
66.5	14.5	210.25
58.5	6.5	42.25
42.5	−9.5	90.25
41.5	−10.5	110.25
49.0	−3.0	9.00
69.0	17.0	289.00
38.5	−13.5	182.25
		$\Sigma(x - \overline{X})^2 = 2966$

3 Now you need to use the following formula. You do **not** need to memorise this formula, because if you are asked to use it in the exam, you will always be given it.

$$s = \sqrt{\frac{\Sigma(x - \overline{X})^2}{n - 1}}$$

where s is the standard deviation
        n is the number of readings (which in this case was 20).

So, $s = \sqrt{\dfrac{2966}{19}} = 12.49$

This should be rounded up to 12.5.

All the values used in the calculations are in mm, so the standard deviation is also in mm. You can say that the mean length of the leaves in this sample is 52.0 mm, with a standard deviation of 12.5. This is often written as: 52.0 mm ± 12.5 mm.

Calculating standard error

The set of data about leaf lengths was taken from a *sample* of leaves. How likely is it that the mean of these data is actually the true mean for the entire population? We can get some idea of this by calculating the standard error for these data. The standard error tells us how far away the actual mean for the entire population might be from the value we have calculated for our sample.

The formula for standard error, S_M, is: $S_M = \dfrac{s}{\sqrt{n}}$

For the leaf data, this works out as: $\dfrac{12.5}{\sqrt{20}} = \dfrac{12.5}{4.5} = 2.8$ mm

This tells us that, if we took a fresh sample of leaves from the same population, we could be 95% confident that the mean length would be within 2 × 2.8 mm of our original mean, which was 52.0 mm.

We could use this to show confidence limits on a graph. Let's say you were trying to find out if the lengths of leaves from a species of tree growing at the edge of a wood were different from the lengths of leaves from the same species growing in the middle of a wood. The first set of data (the ones you have been looking at already) came from the edge of the wood. Here are the results you got when you measured 20 leaves from the middle of the wood.

23.0, 46.5, 52.0, 32.5, 43.0, 50.5, 26.5, 31.5, 54.0, 28.0
47.5, 39.0, 58.5, 47.5, 33.5, 42.5, 26.5, 51.0, 38.0, 38.5

The mean of these values is 40.5 mm.

Try calculating the standard deviation for yourself. You should find that it is 10.4 mm.

Now try calculating the standard error for these data. You should find that it is 2.3 mm. The values for standard error can be shown as error bars on a bar chart.

Each error bar extends 2 × 2.3 mm above and below the plotted value.

The error bars show the 95% confidence limits.

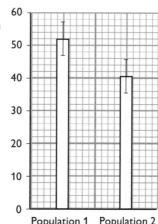

If the values for the two error bars overlap, this means there is not a significant difference between the mean lengths of the two populations of leaves. Here, there is no overlap, so we can say that it is possible that there is a significant difference between the lengths of the leaves from the middle and the edge of the wood.

The chi-squared test

This test is used to tell you whether any difference between your observed and expected values is due to chance, or whether it means that your null hypothesis cannot be correct. It can also be used to test for whether there is an association between two variables — for example, whether people who are left-handed are more likely to study Physics A level than people who are right-handed. The chi-squared test is described on pages 183 to 185.

The t-test

This test is used to tell you whether the means of two sets of values, each following a normal distribution, are significantly different from one another.

We can use the leaf length data to illustrate how this test is done, and what it tells us.

Here are the two sets of data again:

leaves from the edge of the wood
33.5, 56.5, 62.0, 75.0, 36.0, 54.5, 43.5, 41.5, 54.0, 53.0
53.5, 39.0, 72.5, 66.5, 58.5, 42.5, 41.5, 49.0, 69.0, 38.5

leaves from the middle of the wood
23.0, 46.5, 52.0, 32.5, 43.0, 50.5, 26.5, 31.5, 54.0, 28.0
47.5, 39.0, 58.5, 47.5, 33.5, 42.5, 26.5, 51.0, 38.0, 38.5

Just as for the chi-squared test, you begin by constructing a null hypothesis. This would be:

There is no significant difference between the means of the lengths of the leaves from the edge of the wood and from the middle of the wood.

The formula for the t-test will always be given to you, so you don't need to learn it. It is:

$$t = \frac{|\bar{X}_1 - \bar{X}_2|}{\sqrt{\dfrac{s_1^2}{n_1} + \dfrac{s_2^2}{n_2}}}$$

The symbols all mean the same as before. We've already calculated these values. They are:

$\bar{X}_1$ is the mean for the first set of leaves, 52.0 mm

$\bar{X}_2$ is the mean for the second set of leaves, 40.5 mm

s_1 is the standard deviation for the first set of leaves, 12.5 mm

s_2 is the standard deviation for the second set of leaves, 10.4 mm

n_1 and n_2 are the numbers in each sample, which were both 20.

Substituting these values into the formula, we get:

$$t = \frac{|52 - 40.5|}{\sqrt{\dfrac{12.5^2}{20} + \dfrac{10.4^2}{20}}}$$

$$t = \frac{11.5}{\sqrt{\dfrac{156.25}{20} + \dfrac{108.16}{20}}}$$

t = 3.16

What does this mean? You now have to look up this value of t in a table of probabilities.

First, you need to decide how many degrees of freedom there are in your data. To do this, use the formula:

degrees of freedom = $n_1 + n_2 - 2$

$$= 20 + 20 - 2$$

So the number of degrees of freedom is **38**.

Here is a small part of the table showing values of t.

Degrees of freedom	Probability that null hypothesis is correct			
	0.10	0.05	0.02	0.01
25	1.70	2.06	2.49	2.79
30	1.70	2.04	2.46	2.75
40	1.68	2.02	2.42	2.70

Your value of t is 3.16. The table does not have a row for 38 degrees of freedom, so we can look at the closest one, which is 40. There is no value in this row that is as high as your value for t. The values in the table are increasing from left to right, so we can say that our value of t would lie well to the right of anything in the table. This means that the probability of the null hypothesis being correct is much less than 0.01.

Just as we did for the chi-squared test, we take a probability of 0.05 as being the decider for the t-test. If the probability is lower than this, then we can say it is very unlikely that the null hypothesis is correct. In other words, there *is* a significant difference between the mean lengths of the two populations of leaves. The mean length of the leaves in the middle of the wood is significantly less than the mean length of the leaves at the edge of the wood.

Evaluation

This skill builds on what you have already been doing at AS. You need to be able to:

- **spot anomalous data**. These are values that do not fit into a strong pattern that is shown by the rest of the data. This may indicate that you made a mistake when taking a measurement, or that some other variable had changed significantly for that one reading. The best thing to do with genuinely anomalous data is to take them out of your results. Do not include them in calculations of means, and do not take any notice of them when deciding where to draw a line on a graph.

- **consider the need for repeats or replicates**. If you have decided that a data point is anomalous, then it would be a good idea to take repeat readings for that value. Only by repeating those readings can you be sure that your odd result really is anomalous. In any case, it is often a good idea to include a set of repeats or replicates, as explained on pages 108–109 and 116.
- **consider the need for an expansion of the range**, or a change in the interval, of the independent variable.
- **consider major sources of error in the investigation**. These are discussed on pages 115–117. You should be able to judge how much these might have affected the reliability of the results.

Drawing conclusions

Here again, this builds on what you did at AS, where you were also required to draw conclusions from data. Now, though, you may also have statistical analyses to help you to do this. Remember that you will be expected to use the data provided, or that you have calculated, to support your conclusion.

You may also be asked to give biological explanations of the data provided. So it is important that you go into the examination with the facts and concepts in your head that you have learned throughout your AS and A2 course.

Tips During your course:
- Take every opportunity to practise calculating and using statistical tests.
- Make you sure you know how to calculate the number of degrees of freedom.
- The most important part of a statistical test is deciding what you can conclude from it. Make sure you get plenty of practice in doing this during your course.

In the exam:
- If asked, show clearly how you would use tables and graphs to display your data. Show the complete headings for rows, columns and axes, including units.
- Show every step in each calculation that you are asked to do. Even if you get the correct answer, you will not get full marks if steps are missing.
- Be prepared to suggest which would be the best statistical test to use in a particular situation.
- You won't normally be asked to work through an entire statistical test, but only part of it. Read the information you are given very carefully.

A2
Questions
&
Answers

Exemplar paper

This practice examination paper is similar to the CIE A2 Biology Paper 4. All the questions are based on the topic areas described in the previous sections of this book.

You have 2 hours to do the paper. There are 100 marks on the paper, so you can spend just over 1 minute per mark. 85 marks are for structured questions, and there is one 15 mark question that requires more extended writing. You will get a choice of one from two questions in the actual exam paper, but in this sample paper there is only one question.

See page 122 for advice on using this practice paper.

Section A

Question 1

The flow diagram shows part of the metabolic pathway of glycolysis.

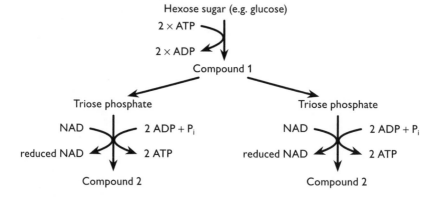

(a) Name compound 1 and compound 2. (2 marks)

(b) State the part of the cell in which this metabolic pathway takes place. (1 mark)

(c) (i) Describe how compound 2 is converted to lactate in a human muscle cell, if oxygen is not available in the cell. (2 marks)

(ii) Describe what happens to the lactate produced. (2 marks)

Total: 7 marks

Candidate A

(a) Compound 1 is fructose phosphate. ✗ Compound 2 is pyruvate. ✓

🖉 Compound 1 should be bisphosphate, not phosphate (it has two phosphate groups attached to it.) 1/2

(b) Cytoplasm. ✓

🖉 Correct. 1/1

(c) (i) This is anaerobic respiration. The pyruvate is changed into lactate so it doesn't stop glycolysis happening.

🖉 The candidate has not answered the question. 0/2

(ii) It goes to the liver, which breaks it down. ✓ This needs oxygen, which is why you breathe faster than usual when you've done a lot of exercise.

🖉 Again, much more is needed for full marks. The comment about breathing rate is correct, but is not relevant to this particular question. 1/2

Candidate B

(a) Compound 1 is hexose bisphosphate. ✓ Compound 2 is pyruvic acid. ✓

🖉 Both correct. 2/2

(b) Cytoplasm. ✓

🖉 Correct. 1/1

(c) (i) It is combined with reduced NAD, ✓ which is oxidised back to ordinary NAD. The enzyme lactate dehydrogenase ✓ makes this happen.

🖉 Entirely correct. 2/2

(ii) The lactate diffuses into the blood and is carried to the liver cells. ✓ They turn it back into pyruvate again, ✓ so if there is oxygen it can go into a mitochondrion and go through the Krebs cycle. Or the liver can turn it into glucose again, ✓ and maybe store it as glycogen.

🖉 All correct and relevant. 2/2

Question 2

The diagram shows the structure of a chloroplast.

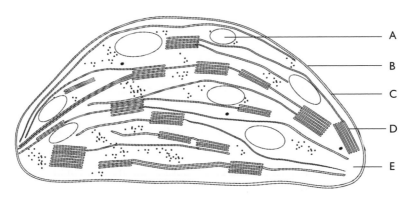

(a) Give the letter of the part of the chloroplast where each of the following takes place.

    (i) fixation of carbon dioxide   (1 mark)

    (ii) the light-dependent reactions   (1 mark)

(b) A grass adapted for growing in a tropical climate was exposed to low temperatures for several days. The membranes of part D moved closer together, so that there was no longer any space between them. This prevented photophosphorylation taking place.

Explain how this would prevent the plant from synthesising carbohydrates.   (3 marks)

(c) Two groups of seedlings were grown in identical conditions for 2 weeks. One group was then grown in high-intensity light and the other group in low-intensity light, for 4 weeks.

Each group of plants was then placed in containers in which carbon dioxide concentration was not a limiting factor. They were exposed to light of varying intensities and their rate of carbon dioxide uptake per unit of leaf area was measured.

The results are shown in the graph.

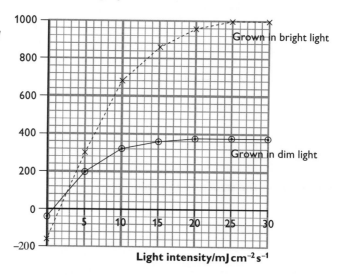

    (i) Compare the effect of light intensity on the two groups of plants between 0 and $30\,\mathrm{mJ\,cm^{-2}\,s^{-1}}$.   (4 marks)

    (ii) Suggest why the plants gave out carbon dioxide at very low light intensities.   (2 marks)

    (iii) Suggest two differences in the two groups of plants that could have been caused by their exposure to different intensities of light

as they were growing, and that could help to explain the results shown in the graph. (2 marks)

Total: 13 marks

Candidate A

(a) (i) E ✓   **(ii)** D ✓

    ✎   Both correct. 2/2

(b) It would not be able to make any ATP, ✓ which is needed for the Calvin cycle. ✓ Without the Calvin cycle, it would not be able to make carbohydrates.

    ✎   This is correct, but lacking in detail. 2/3

(c) (i) Neither of the groups took up any carbon dioxide when there was no light. ✓ Then the quantity of carbon dioxide increased dramatically for the bright light plants, and slowly for the dim light plants. Then it levelled out, lower for the dim light plants than for the bright light ones. ✓ The dim light plants levelled out at 380 and the bright light ones at $1000\,\mu mol\,dm^{-2}\,h^{-1}$. ✓

    ✎   There are some good comparative points made here. A fundamental error, however, is that the candidate expresses his or her answer as though the x axis showed time — for example using the word 'slowly' and 'then'. It is also not a good idea to use terms such as 'dramatically'. See candidate B for a better way of expressing these points. 3/5

(ii) They could not photosynthesise, ✓ so the carbon dioxide in their leaves just went back out into the air again.

    ✎   One correct point is made here. 1/2

(iii) The ones that had grown in the bright light could have bigger leaves and more chlorophyll. ✓

    ✎   The suggestion about bigger leaves is not correct. Even if the plants did have bigger leaves, this would not affect the results, because the carbon dioxide uptake is measured per unit area (look at the units on the y axis of the graph). The second point is a good suggestion. 1/2

Candidate B

(a) (i) E ✓   **(ii)** D ✓

    ✎   Both correct. 2/2

(b) No ATP would be made ✓ in the light-dependent reaction, so there would not be any available for the light-independent reactions, where carbohydrates (triose phosphate) are made in the Calvin cycle ✓. ATP is needed to convert GP to triose phosphate ✓ (along with reduced NADP) and also to help regenerate RuBP ✓ from the triose phosphate so the cycle can continue.

    ✎   All correct and with good detail. 3/3

(c) (i) Below about 0.5 light intensity, both groups gave out carbon dioxide. ✓ As light intensity increased, the amount of carbon dioxide taken up by the plants grown in bright light increased more steeply ✓ than for the ones grown in dim light. In the group grown in dim light, the maximum rate of carbon dioxide uptake was $380\,\mu mol\,dm^{-2}h^{-1}$, whereas for the ones in bright light it was much higher, ✓ at $1000\,\mu mol\,dm^{-2}h^{-1}$. ✓ For the bright light plants, the maximum rate of photosynthesis was not reached until the light intensity was $25\,mJ\,cm^{-2}s^{-1}$, but for the ones grown in dim light the maximum rate was reached at a lower ✓ light intensity of $20\,mJ\,cm^{-2}s^{-1}$.

> ☑ A good answer, with some comparative figures quoted (with their units). Note the avoidance of any vocabulary that could be associated with time. 5/5

(ii) When the light intensity was very low, the plants would not be able to photosynthesise so they would not take up any carbon dioxide. ✓ However, they would still be respiring (they respire all the time) so their leaf cells would be producing carbon dioxide ✓ which would diffuse out into the air. Normally, this carbon dioxide would be taken up by the cells for photosynthesis.

> ☑ A good answer. 2/2

(iii) The plants grown in the light would probably be a darker green because they would have more chlorophyll ✓ in their chloroplasts, so they would be able to absorb more light and photosynthesise faster. They might also have more chloroplasts in each palisade cell. ✓ And leaves sometimes produce an extra layer of palisade cells if they are in bright light. ✓

> ☑ This answer actually contains three points, and two of them have been explained, which was not required. The candidate could have got 2 marks with a much shorter answer. Nevertheless, this answer shows good understanding of the underlying biology. 2/2

Question 3

A group of islands contains 3 species of mice, each species being found on only one island. A fourth species is found on the mainland. A region of the DNA of each species was sequenced, and the percentage differences between the samples were calculated. The results are shown in the table.

	Mainland	Island A	Island B
Island A	6.1		
Island B	4.8	9.7	
Island C	5.2	10.3	7.5

(a) Discuss how these results suggest that the species of mouse on each island has evolved from the species on the mainland, and not from one of the other island species. (5 marks)

(b) Each of these four species of mice is unable to breed with any of the other species, even if they are placed together.

Suggest how reproductive isolation between the mice could have arisen, and explain its role in speciation. (5 marks)

Total: 10 marks

Candidate A

(a) The three island mice each have DNA more similar to the mainland mouse than to each other. ✓✓ So they have probably all evolved from the mainland mouse. If one of the island mice had evolved from another island mouse, their DNA would be more similar. ✓

> ⎙ This answer shows that the candidate has managed to work out what the table shows, but it does not provide enough detail in the discussion to get all of the marks available. 3/5

(b) If they are on different islands, they will have different selection pressures ✓ so the mice might end up different. They might have different courtship behaviour, ✓ so they won't be able to mate with each other. ✓ You have to get reproductive isolation to produce a new species.

> ⎙ A reasonable description. The last sentence is moving towards another mark, but it really only repeats what is already in the question. 3/5

Candidate B

(a) From the table, we can see that each island mouse's DNA is more similar to the DNA of the mainland mouse than to any of the other island mice. ✓ For example, the island C mice have DNA that is 10.3% different from the island A mice and 7.5% different from the island B mice, but only 5.2% different from the mainland mice. ✓ The longer ago two species split away from each other, the more different we would expect their DNA to be. ✓ This is because the longer the time, the more mutations ✓ are likely to have occurred, so there will be different base sequences ✓ in the DNA.

> ⎙ A good answer to a difficult question. The candidate has made good use of the data in the table, and has used some of the figures to support his or her answer. The last part of the answer explains why differences in DNA base sequence indicate degree of relationship. 5/5

(b) A species is defined as a group of organisms that can interbreed with each other to produce fertile offspring. ✓ So to get new species you have to have something that stops them reproducing together so genes can't flow ✓ from one species to the other. This might happen if different selection pressures ✓ acted on two populations of a species, so that different alleles were selected for ✓ and over many generations their genomes became more different. ✓ So they might be the wrong size and shape ✓ to be able to breed with each other.

⏚ The answer begins with a clearly explained link between speciation and reproductive isolation, and then goes on to describe how two populations could
become reproductively isolated. 5/5

Question 4

**The diagram shows a section through a wheat grain. Wheat, like maize,
is a cereal plant.**

(a) On the diagram, use a label line to label and name each of the
following structures.

endosperm	cotyledon
embryo plant	fruit wall and testa

(2 marks)

(b) Explain the significance of the grains of cereal crops in the human
diet. (3 marks)

(c) Explain how gibberellin is involved in the germination of a wheat
seed. (5 marks)

Total: 10 marks

Candidate A

(a)

⏚ The cotyledon label is incorrect. 1/2

(b) Cereals are grasses that are cultivated for their seeds. They are staple foods
in many parts of the world, because they are easy to grow and harvest. Rice,
wheat, barley and maize are all cereal crops. They have a lot of starch ✓ in their
endosperms.

⏚ Everything that the candidate has written is correct, but most of it does not
answer the question, which asks about the role *in the diet*. 1/3

(c) Gibberellin (GA) is made by the seed when it is warm and it takes up water. ✓
The GA breaks the dormancy of the seed. The GA makes the cells in the seed

secrete amylase, ✓ which breaks down the starch stores so the embryo can use them to grow.

🖉 Once again, the candidate has written nothing wrong, but more detail is expected at this level. For example, which parts of the seed secrete the GA and the amylase? 2/5

Candidate B

(a)

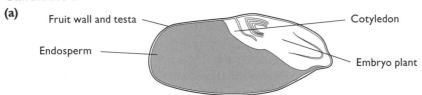

🖉 All correct. 2/2

(b) The endosperm of cereal grains is rich in starch, ✓ which is stored for the use of the embryo at germination. This is also excellent food for humans, as it is high in energy. Cereal grains also contain significant amounts of protein, mostly from the aleurone layer. ✓ Cereal grains can be stored for long periods of time, which makes them useful as components of the diet even at parts of the year when the crops are not growing. ✓

🖉 A good answer, making three good points. The candidate explains not only what nutrients the cereal grains contain, but also which part of the seed contains them and also how they contribute to a human diet. The comment about storage is also relevant to this question. 3/3

(c) When a seed takes up water, this stimulates the synthesis of GA ✓ by the tissues of the embryo plant. ✓ The GA in turn stimulates the synthesis of amylase ✓ by the aleurone layer. ✓ The amylase hydrolyses starch stored in the endosperm, ✓ producing maltose, which is soluble and is transported to the embryo for use as an energy resource ✓ and also as raw material from which cellulose can be synthesised to make new cell walls. ✓

🖉 This answer has plenty of relevant detail, and scores the maximum possible number of marks. The candidate tells us what stimulates the production of GA, where it is made and then what it does. We are also told why this is important. 5/5

Question 5

The Irish Threatened Plant Genebank was set up in 1994 with the aim of collecting and storing seeds from Ireland's rare and endangered plant species. The natural habitat of many of these species is under threat. For each species represented in the bank, seeds are separated into active and base collections. The active collection contains seeds that are available for immediate use, which could be for reintroduction into the wild, or for germination to produce new plants and therefore new seeds. The base collection is left untouched. Some of the base collection is kept in Ireland, and some is kept at seed banks in other parts of the world.

(a) Suggest why seed banks separate stored seeds into active and base collections.
(2 marks)

In 2001, an investigation was carried out into the effect of long-term storage on the ability of the seeds to germinate. Fifteen species were tested. In each case, 100 seeds were tested. It was not possible to use more because in many cases this was the largest number that could be spared from the seed bank. In most cases, the germination rate of the seeds had already been tested when they were first collected in 1994, so a comparison was possible with the germination rates in 2001 after 7 years of storage.

The graphs show the results for two species, *Asparagus officinalis* and *Sanguisorba officinalis*.

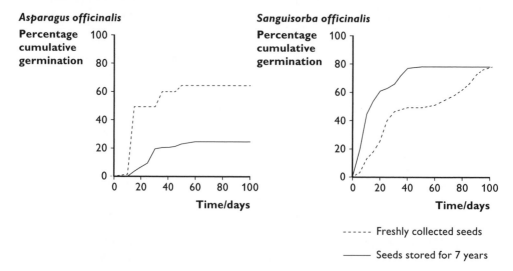

(b) (i) Compare the germination rates of stored and fresh seeds of *Sanguisorba officinalis*.
(3 marks)

(ii) Compare the effect of storage on the germination rates of *Sanguisorba officinalis* and *Asparagus officinalis*.
(3 marks)

(c) It has been suggested that species stored as seeds in seed banks have different selection pressures acting on them compared with the same species living in the wild.

(i) Explain why the selection pressures in a seed bank and in the wild are likely to be different. (2 marks)

(ii) Suggest how the possible harmful effects of these differences could be minimised. (5 marks)

Total: 15 marks

Candidate A

(a) So they always have some spare.

🖉 Not enough for a mark. 0/2

(b) (i) The stored ones germinated better than the fresh ones. The stored ones go up more quickly than the fresh ones. But they all end up at the same place, about 80%. ✓

🖉 This answer loses out by poor wording, and not being clear enough about exactly what is being described. The word 'better' in the first sentence could mean that the seeds germinated more quickly, or that more of them germinated, so the candidate needs to clarify this. 'Go up more quickly' is also not clearly related to germination. The last sentence is quite generously given a mark for the idea that eventually about 80% of the seeds in each batch germinated. 1/3

(ii) The *Sanguisorba* seeds germinated better when they had been stored, but the *Asparagus* seeds germinated better when they were fresh. Storing the *Sanguisorba* seeds made them germinate faster, but the *Asparagus* seeds germinated slower. ✓

🖉 Again, poor wording means that this answer only gets 1 mark. The word 'better' is used in the first sentence and, as has been explained above, this is not a good word to use in this context. One mark is given for the idea that storage caused slower germination in *Asparagus* but faster germination in *Sanguisorba*. 1/3

(c) (i) The conditions in which the seeds grow might be different in the seed bank from in the wild.

🖉 This is not quite correct as seeds do not grow. Growth only happens after germination, so it is seedlings and plants that grow. The candidate needs to think more carefully about what happens to seeds in a seed bank. 0/2

(ii) The seeds could be grown in conditions like those in the wild.

🖉 Again, this hasn't been clearly thought out. 0/5

Candidate B

(a) Having active collections is good because it means there are seeds available that can be used for something. But you must always keep some seeds in storage, because the whole point of a seed bank is that it stores seeds and these need to be kept safe so they don't get destroyed. ✓ If all of them got used for growing plants, then perhaps the plants would die ✓ and you wouldn't have any seeds left.

🖉 This is not very well expressed, but the right ideas are there. 2/2

(b) (i) The stored seeds germinated much faster ✓ than the fresh ones. By 10 days, about 60% of the stored seeds had germinated, but only about 8% ✓ of the fresh ones. By 50 days, all of the stored seeds that were going to germinate (about 80%) had germinated. It took 100 days for 80% of the fresh seeds to germinate. ✓ We can't tell if any more would have germinated after that because the line is still going up when the graph stops. ✓

🖉 Clear comparative points have been made about the speed at which germination happened, and also about the maximum percentage of seeds that germinated. 3/3

(ii) Storage seemed to help the germination for *Sanguisorba*, but it made it worse for *Asparagus*. For *Asparagus*, storage made it germinate slower, but it germinated faster for *Sanguisorba*. ✓ And for *Asparagus* only about 30% of the stored seeds germinated compared with 60% of the fresh seeds, ✓ but with *Sanguisorba* about the same percentage of seeds germinated for both fresh and stored. ✓

🖉 A good answer, which again makes clear comparisons and discusses both the speed of germination and the percentage of seeds that eventually germinated. 3/3

(c) (i) In the seed bank, the seeds are just stored. So the ones that survive are the ones that are best at surviving in those conditions as dormant seeds. ✓ In the wild, the plants have to be adapted to grow in their habitat, ✓ so maybe they have to have long roots or big leaves or whatever. In the seed bank, that doesn't matter.

🖉 This is a good answer which really does answer the question, but it could perhaps have been written a little bit more carefully and kept shorter. 2/2

(ii) You might get seeds that are really good at surviving in a seed bank but when they germinate they produce plants that aren't very good at surviving in the wild. To avoid this, you could keep on collecting fresh seeds from plants in the wild, ✓ and only storing them for a little while before replacing them with new ones. ✓ If you had to store seeds for a long time, you could keep germinating some of them ✓ and growing them in conditions like in the wild ✓ and then collect fresh seeds from the ones that grew best. ✓

🖉 This is a really good answer to a tricky question. The candidate has made several sensible suggestions, including storing the seeds for a shorter time and periodically exposing the plants to natural conditions where the 'normal' selection pressures will operate. 5/5

Question 6

Two parents with normal skin and hair colouring had six children, of whom three were albino. Albino people have no colouring in their skin or hair, due to having an inactive form of the enzyme tyrosinase. Tyrosinase is essential for the formation of the brown pigment melanin.

(a) The normal allele of the tyrosinase gene is **A**, and the allele that produces faulty tyrosinase is a.

   State the genotypes of the parents and their albino children.　　(2 marks)

(b) Albinism is a relatively frequent condition in humans, but one of these albino children had a very unusual phenotype. While most of her hair was white, the hair of her eyebrows developed some brown colouring, as did the hair on her hands and lower legs. Genetic analysis suggested that a mutation had occurred in the faulty tyrosinase allele.

   Suggest why it is likely that this mutation occurred in the ovaries or testes of the girl's parents, rather than in her own body.　　(2 marks)

(c) The graph below shows how the activity of normal tyrosinase and tyrosinase taken from the albino girl were affected by temperature.

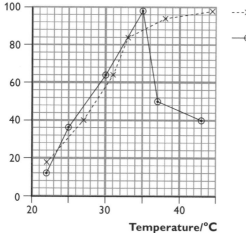

(i) Compare the effects of temperature on normal tyrosinase and the albino girl's tyrosinase.　　(3 marks)

(ii) Studies on the production and activity of tyrosinase in living cells found that normal tyrosinase leaves the endoplasmic reticulum shortly after it has been made, and accumulates in vesicles in the cytoplasm where it becomes active. However, the tyrosinase in the albino girl's cells only did this at temperatures of 31 °C or below. **At 37°C, her tyrosinase accumulated in the endoplasmic reticulum.**

Use your answer to (i) and the information above to discuss explanations for the distribution of colour in the hair on different parts of the albino girl's body. (4 marks)

Total: 11 marks

Candidate A

(a) Aa and aa ✓✓

> 🖉 The candidate has not made clear which genotype refers to the parents and which to the children, but has generously been given the benefit of doubt. 2/2

(b) If it had been in one of her own cells, then only that cell would be affected and not the whole body. ✓

> 🖉 Correct as far as it goes. 1/2

(c) (i) Both enzymes get more active as temperature increases. ✓ The normal enzyme is still getting more active even at 45°C. ✓ The girl's enzyme has an optimum activity at 35°C ✓ and above that it quickly ✗ gets a lot less active.

> 🖉 The answer begins well with a sentence about both enzymes. Two clear distinguishing points are then made. However, the use of the term 'quickly' is inappropriate as there is nothing about time on the graph. 'Steeply' would have been better. 3/3

(ii) Maybe her legs and eyebrows and hands were colder, ✓ like in a Himalayan rabbit. So the enzyme would not work well where it was hotter ✓ and would not make melanin.

> 🖉 This is a difficult question, and the candidate has done well to think of what he/she had learned about the interaction between genes and environment in animal hair colour and recognise that it could relate to this situation. However, the answer does not refer clearly to the information provided, and this needs to be done in order to achieve more marks. 2/4

Candidate B

(a) Parents Aa and Aa. ✓ Albino children aa. ✓

> 🖉 Correct and clear. 2/2

(b) If it had occurred in her own body, then it would probably be in only one cell ✓ so you wouldn't see the effects in cells in different parts of the body. If it had been in one of the parents, then the mutated gene would have been in the zygote, ✓ so when the zygote divided by mitosis the gene would get copied ✓ into every cell in her body.

> 🖉 Absolutely correct. 2/2

(c) (i) The activity of both enzymes increases as temperature increases from 22°C to 35°C. ✓ However, above that the activity of the girl's enzyme drops really steeply ✓ so it has only 50% activity at 37°C. But the normal enzyme keeps on increasing its activity up to where the graph ends at 45 °C. ✓ At 38 °C it has 97% activity, almost twice as much as for the girl's enzyme. ✓

🖉 A good answer which mentions particular points on the curves and provides a quantitative comparison between the figures for the two enzymes at a particular temperature (38 °C). 3/3

(ii) The extremities of the girl's body will be cooler than other parts, ✓ so the tyrosinase will be able to work here because according to the graph it can work well up to 35 °C. ✓ It will also be able get out of the endoplasmic reticulum ✓ in her cells and into the vesicles where it becomes active. This is why the hairs on her hands and legs are darker because the enzyme was able to make melanin there. ✓ But in warmer places the enzyme stops working and can't get out of the RER so no melanin is made. ✓

🖉 An excellent answer. The candidate uses both sets of information (from the graph and the description about the endoplasmic reticulum) to provide an explanation linked to what he/she had already learned about the control of hair colour in other situations. 4/4

Question 7

(a) **Describe how gel electrophoresis can be used to separate DNA fragments of different lengths.** (6 marks)

(b) **The diagram shows the results of electrophoresis on DNA samples taken from a mother, her child and its alleged father.**

Explain what can be concluded from these results. (3 marks)

Total: 9 marks

Candidate A

(a) First you cut the DNA up into pieces using restriction enzymes. ✓ Then you put the DNA onto some agarose gel ✓ in a tank and switch on the power supply so the DNA gets pulled along the gel. ✓ The bigger pieces move more slowly, so they end up not so far along the gel as the smaller pieces. ✓ You can't see the DNA so you need to stain it with something so it shows up.

⚘ There are no errors in this answer, but a little more detail is needed. 4/6

(b) The child has one band that is in its mother's DNA and another that is in the alleged father's DNA. ✓ So he could be the father. ✓

⚘ Two correct points made. 2/3

Candidate B

(a) The DNA is cut into fragments using restriction enzymes, ✓ which cut it at particular base sequences. Then you place samples of the DNA into little wells in agarose gel ✓ in an electrophoresis tank. A voltage is then applied ✓ across the gel. The DNA pieces have a small negative charge so they steadily move towards the positive ✓ terminal. The larger they are, the more slowly they move ✓ so the smaller ones travel further ✓ than the big ones. After a time, the power is switched off so the DNA stops moving. You can tell where it is by using radio-activity, ✓ so the DNA shows up as bands on a photographic film.

⚘ All correct and enough detail for full marks. 6/6

(b) The top band for the child matches the top band for the mother, ✓ and the bottom band for the child matches the bottom band for the father. ✓ So the alleged father could be the child's father, ✓ though we can't be 100% certain of that because there could be another man who has this band as well. ✓

⚘ A clear and thorough conclusion. 3/3

Question 8

(a) The diagram shows a section through part of a human ovary just before ovulation.

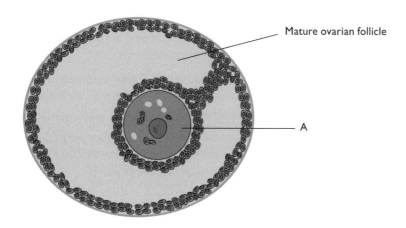

Mature ovarian follicle

A

(i) Identify structure A.   (1 mark)

**(ii) Describe what will happen to the follicle at ovulation and in the
          next few days.**   (3 marks)

**(b) A woman can find out if she is pregnant by testing her urine for the
       presence of human chorionic gonadotropin, hCG. Modern pregnancy
       testing kits use monoclonal antibodies to test for the presence of this
       hormone.**

(i) Explain the meaning of the term *monoclonal antibody*.   (2 marks)

(ii) Describe how monoclonal antibodies are used to test for hCG.   (4 marks)

Total: 10 marks

Candidate A

(a) (i)   ovum

> 🖉   Not correct. In humans, meiosis is not completed until after the egg has left
>   the ovary and has been fertilised, so the egg is still a secondary oocyte at this
>   stage. 0/1

(ii) The follicle will burst and the egg will shoot out into the funnel of the
          oviduct. ✓ It will travel down the oviduct where it might be fertilised. The
          follicle will close back up again and turn into a corpus luteum. ✓

> 🖉   All correct. However, there is not enough information about what happens
>   to the follicle after ovulation. The information about what happens to the *egg*
>   (the second sentence of the answer) is not relevant to this question and the
>   candidate has wasted time and space with this part of the answer. 2/3

(b) (i)   It's antibodies that are made by injecting mice with antigens and then
              fusing their plasma cells with cancer cells that make lots of the same kind
              of antibody. ✓

🖉 The candidate has implied that monoclonal antibodies are many antibodies all of exactly the same type. However, the answer only explains the meaning of the term 'monoclonal', and not the term 'antibody'. It's important to explain *both* words in a two-word term.1/2

(ii) The antibodies are stuck to hCG antibodies ✔ that are stuck to a little piece of plastic that you dip into urine. If there is hCG in the urine it sticks to the antibodies ✔ which change colour.

🖉 This answer gives a very general idea of how the procedure works. The candidate knows that the kit involves the use of antibodies to hCG, and that the binding of these with hCG results in a colour change. 2/4

Candidate B

(a) (i) secondary oocyte ✔

🖉 Correct. 1/1

(ii) The ovarian follicle is at the edge of the ovary, so when it bursts the secondary oocyte goes out of the ovary and into the funnel of the oviduct. ✔ The space inside the follicle now fills up with yellow cells (from the granulosa cells round the follicle) that form a structure called a corpus luteum. ✔ These cells secrete the hormone progesterone, ✔ which helps to maintain the lining of the uterus ready for implantation if the egg is fertilised.

🖉 An entirely correct, complete and relevant answer. 3/3

(b) (i) They are antibodies that are all of exactly the same kind, that will all bind to the same substance. ✔ An antibody is an immunoglobulin (protein) that binds to a particular antigen. ✔ They are made by injecting a mouse or other animal with an antigen, which stimulates certain lymphocytes to divide by mitosis to form a clone of plasma cells which all secrete the antibody against this antigen. The plasma cells are fused with cancer cells to form hybridomas, which divide to form large colonies of cells all secreting the antibody.

🖉 The candidate answered the question perfectly in the first two sentences, and all the rest is unnecessary, as it describes how monoclonal antibodies are made and not what they are. 2/2

(ii) Monoclonal antibodies that will bind to hCG are fixed to a dipstick (immobilised). ✔ When the stick is placed in urine, any hCG binds to the antibodies. ✔ The bound antibodies are carried up the stick as the urine seeps upwards. They reach an area where there are other antibodies that will only bind with antibodies that have already bound with hCG. ✔ When they do this, that area changes colour. ✔

🖉 This is a correct answer, and there is just enough detail to get all four marks. (Note that there are other types of dipstick that work in slightly different ways — it does not matter if you have learned about a different one as the examiners will credit any correct description.) 4/4

Section B
Question 9

(a) Describe how a nerve impulse crosses a cholinergic synapse. (9 marks)

(b) Outline the functions of a sensory neurone and a motor neurone in a reflex arc. (6 marks)

Total: 15 marks

Candidate A

(a) In order for the action potential to cross the synaptic cleft, it has to go through a process.

A nerve impulse arrives at a synapse in the form of an action potential. It causes sodium and calcium gates to open so these ions ✓ go into the presynaptic membrane. ✗ The calcium causes the vesicles in the presynaptic neurone to fuse with the membrane ✓ and release acetycholine into the synapse. This binds to the receptors on the post synaptic membrane ✓ and this makes them pump ✗ sodium ions into the neurone which starts up another action potential in the second neurone, because it depolarises it. ✓

Then the enzyme acetylcholinesterase causes the acetylcholine to go back into the presynaptic neurone, ✗ so it returns to the resting potential again.

> 🖉 This answer is not very clear, and the examiner cannot always be certain what the candidate means. For example, the calcium ions go through the presynaptic membrane and into the neurone, not 'into' the membrane. Notice that no mark was given for the mention of calcium until the candidate used the term 'ions' later on. There is an error in the next to last sentence; the sodium ions are not pumped into the postsynaptic neurone, but pass through the opened sodium ion channels by diffusion, down their concentration gradient. The last sentence also contains an error, as acetycholinesterase breaks down acetylcholine. In any case, this process is not strictly relevant to a description of how the impulse crosses the synapse, because it takes place after that event has finished. Overall, not a strong answer, with lack of detail and several errors. 4/9

(b) The sensory neurone transmits an action potential from a receptor into the spinal cord. ✓ The action potential crosses a synapse and then passes along a motor neurone to an effector, such as muscle. ✓ This makes the muscle contract. ✓ This is a reflex action. The muscle automatically contracts without you having to think about it.

> 🖉 There are three correct statements here, but all of this could have been answered by a good IGCSE candidate, and there is not sufficient detail for a high mark at A level. 3/6

Candidate B

(a) When the action potential arrives at the presynaptic knob, ✓ it causes calcium ion channels to open. ✓ Ca^{2+} flood into the neurone, ✓ down their concentration gradient. ✓ The knob contains many tiny vesicles full of the neurotransmitter, ✓ acetylcholine. ✓ The calcium ions make these vesicles move to the presynaptic membrane and fuse with it, ✓ releasing the acetylcholine into the synaptic cleft. ✓

This cleft is very small, so it takes only a millisecond or two for the acetylcholine to diffuse ✓ across it. On the other side of the cleft, there are receptor molecules in the postsynaptic membrane, and the acetylcholine molecules fit perfectly into these. ✓ This makes sodium ion channels in the postsynaptic membrane open, ✓ so sodium ions flood in down their concentration gradient. ✓ This depolarises the membrane ✓ (gives it a positive charge inside) which sets up an action potential in the postsynaptic neurone.

> ✍ This is a good answer. Although it is short, it is packed with correct and relevant detail. There are no mistakes, and no important steps have been omitted. It is usual for the Section B questions to have many more marking points than the total number of marks available, so even though this answer has 13 ticks it can still only get the maximum 9/9.

(b) A sensory neurone has its cell body in the ganglion in the dorsal root of a spinal nerve. It has a very long dendron that carries action potentials from a receptor towards its cell body, ✓ and a shorter axon that carries the action potentials into the spinal cord (or brain). ✓ The ending of the dendron may be within a specialised receptor such as a Pacinian corpuscle in the skin. ✓ Pressure acting on the Pacinian corpuscle depolarises the membrane of the dendron and generates an action potential. ✓ (In general, receptors transfer energy from a stimulus into energy in an action potential.)

The motor neurone has its cell body within the central nervous system (in the brain or the spinal cord). It has many short dendrites and a long axon. It will have many synapses, including several with sensory neurones. ✓ Thus the action potential from a sensory neurone can cross the synapse and set up an action potential in the motor neurone, ✓ which will then transmit it to an effector such as a muscle or gland. ✓ The action potential then causes the effector to respond, for example by contracting (if it is a muscle). ✓

In a reflex arc, the impulses travel directly from the sensory to the motor neurone (or sometimes via an intermediate neurone between them) without having to be processed in the brain. ✓ This means the pathway from receptor to effector is as short as possible, so the response can happen very quickly.

> ✍ A good answer, taking each neurone in turn and concentrating on functions with only brief references to structure where these are directly relevant to function. 6/6